Leveled Texts
for Science
Earth and Space Science

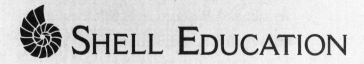
SHELL EDUCATION

These resources were purchased with Title III funds

Reading Level Editor
Josh BishopRoby

English Language Learner Consultants
D. Kyle Shuler
Chino Valley Unified School District, California
Marcela von Vacano
Arlington County Schools, Virginia

Gifted Education Consultant
Wendy Conklin, M.A.
Mentis Online, Round Rock, Texas

Special Education Consultant
Dennis Benjamin
Prince William County Public Schools, Virginia

Contributing Content Authors
Gina dal Fuoco
Connie Jankowski
Lisa E. Greathouse
Greg Young
Lynn Van Gorp, M.S.
William B. Rice
Debra J. Housel

Publisher
Corinne Burton, M.A.Ed.

Editorial Director
Dona Herweck Rice

Creative Director
Lee Aucoin

Editor-in-Chief
Sharon Coan, M.S.Ed.

Editorial Manager
Gisela Lee, M.A.

Cover Art
Lesley Palmer

Print Production
Neri Garcia

Shell Education
5301 Oceanus Drive
Huntington Beach, CA 92649
http://www.shelleducation.com
ISBN 978-1-4258-0160-1

© 2008 Shell Education
Reprinted 2009

Table of Contents

Introduction 4
What Is Differentiation? 4
How to Differentiate Using This Product 5
General Information About the Student Populations 6
    Special Education Students 6
    English Language Learners 6
    Regular Education Students 7
    Gifted Education Students 7
Strategies for Using the Leveled Texts 8
    Special Education Students 8
    English Language Learners 11
    Gifted Education Students 14
How to Use This Product 16
    Readability Chart 16
    Components of the Product 16
    Tips for Managing the Product 18
    Correlation to Standards 19
Leveled Texts
    Jet Streams and Trade Winds 21
    The Water Cycle 29
    Tornadoes and Hurricanes 37
    Structure of the Earth 45
    Earthquakes and Volcanoes 53
    Plate Tectonics 61
    Wegener Solves a Puzzle 69
    The Rock Cycle 77
    Fun with Fossils 85
    The Inner Planets 93
    The Outer Planets 101
    Our Place in Space 109
    Other Citizens of the Solar System 117
    The Astronomer's Toolbox 125
    The Journey to Space 133
Appendix
    Resources 141
        Works Cited 141
        Image Sources 141–143
        Contents of Teacher Resource CD 144

What Is Differentiation?

Over the past few years, classrooms have evolved into diverse pools of learners. Gifted students, English language learners, learning-disabled students, high achievers, underachievers, and average students all come together to learn from one teacher. The teacher is expected to meet their diverse needs in one classroom. It brings back memories of the one-room schoolhouse during early American history. Not too long ago, lessons were designed to be one size fits all. It was thought that students in the same grade level learned in similar ways. Today, we know that viewpoint to be faulty. Students have differing learning styles, come from different cultures, experience a variety of emotions, and have varied interests. For each subject, they also differ in academic readiness. At times, the challenges teachers face can be overwhelming, as they struggle to figure out how to create learning environments that address the differences they find in their students.

What is differentiation? Carol Ann Tomlinson at the University of Virginia says, "Differentiation is simply a teacher attending to the learning needs of a particular student or small group of students, rather than teaching a class as though all individuals in it were basically alike" (2000). Differentiation can be carried out by any teacher who keeps the learners at the forefront of his or her instruction. The effective teacher asks, "What am I going to do to shape instruction to meet the needs of all my learners?" One method or methodology will not reach all students.

Differentiation encompasses what is taught, how it is taught, and the products students create to show what they have learned. When differentiating curriculum, teachers become the organizers of learning opportunities within the classroom environment. These categories are often referred to as content, process, and product.

- **Content:** Differentiating the content means to put more depth into the curriculum through organizing the curriculum concepts and structure of knowledge.

- **Process:** Differentiating the process requires the use of varied instructional techniques and materials to enhance the learning of students.

- **Product:** When products are differentiated, cognitive development and the students' abilities to express themselves improves.

Teachers should differentiate content, process, and product according to students' characteristics. These characteristics include students' readiness, learning styles, and interests.

- **Readiness:** If a learning experience aligns closely with students' previous skills and understanding of a topic, they will learn better.

- **Learning styles:** Teachers should create assignments that allow students to complete work according to their personal preferences and styles.

- **Interests:** If a topic sparks excitement in the learners, then students will become involved in learning and better remember what is taught.

How to Differentiate Using This Product

The leveled texts in this series help teachers differentiate science content for their students. Each book has 15 topics, and each topic has a text written at four different reading levels. (See page 17 for more information.) These texts are written at a variety of reading levels, but all the levels remain strong in presenting the science content and vocabulary. Teachers can focus on the same content standard or objective for the whole class, but individual students can access the content at their instructional levels rather than at their frustration levels.

Determining your students' instructional reading levels is the first step in the process. It is important to assess their reading abilities often so they do not get tracked into one level. Below are suggested ways to use this resource, as well as other resources in your building, to determine students' reading levels.

- **Running records:** While your class is doing independent work, pull your below-grade-level students aside, one at a time. Have them read aloud the lowest level of a text (the star level) individually as you record any errors they make on your own copy of the text. If students read accurately and fluently and comprehend the material, move them up to the next level and repeat the process. Following the reading, ask comprehension questions to assess their understanding of the material. Assess their accuracy and fluency, mark the words they say incorrectly, and listen for fluent reading. Use your judgment to determine whether students seem frustrated as they read. As a general guideline, students reading below 90% accuracy are likely to feel frustrated as they read. There are also a variety of published reading assessment tools that can be used to assess students' reading levels with the running record format.

- **Refer to other resources:** Other ways to determine instructional reading levels include checking your students' Individualized Education Plans, asking the school's ELL and special education teachers, or reviewing test scores. All of these resources should be able to give you the further information you need to determine at which reading level to begin your students.

Teachers can also use the texts in this series to scaffold the content for their students. At the beginning of the year, students at the lowest reading levels may need focused teacher guidance. As the year progresses, teachers can begin giving students multiple levels of the same text to allow them to work independently to improve their comprehension. This means each student would have a copy of the text at his or her independent reading level and instructional reading level. As students read the instructional-level texts, they can use the lower texts to better understand the difficult vocabulary. By scaffolding the content in this way, teachers can support students as they move up through the reading levels. This will encourage students to work with texts that are closer to the grade level at which they will be tested.

General Information About the Student Populations

Special Education Students

By Dennis Benjamin

Gone are the days of a separate special education curriculum. Federal government regulations require that special education students have access to the general education curriculum. For the vast majority of special education students today, their Individualized Education Plans (IEPs) contain current and targeted performance levels but few short-term content objectives. In other words, the special education students are required to learn the same content as their regular education peers.

Be well aware of the accommodations and modifications written in students' IEPs. Use them in your teaching and assessment so they become routine. If you hold high expectations of success for all of your students, their efforts and performances will rise as well. Remember the root word of disability is ability. Go to the root needs of the learner and apply good teaching. The results will astound and please both of you.

English Language Learners

By Marcela von Vacano

Many school districts have chosen the inclusion model to integrate English language learners (ELLs) into mainstream classrooms. This model has its benefits as well as its drawbacks. One benefit is that English language learners may be able to learn from their peers by hearing and using English more frequently. One drawback is that these second-language learners cannot understand academic language and concepts without special instruction. They need sheltered instruction to take the first steps toward mastering English. In an inclusion classroom, the teacher may not have the time or necessary training to provide specialized instruction for these learners.

Acquiring a second language is a lengthy process that integrates listening, speaking, reading, and writing. Students who are newcomers to the English language are not able to process information until they have mastered a certain number of structures and vocabulary words. Students may learn social language in one or two years. However, academic language takes up to eight years for most students.

Teaching academic language requires good planning and effective implementation. Pacing, or the rate at which information is presented, is another important component in this process. English language learners need to hear the same word in context several times, and they need to practice structures to internalize the words. Reviewing and summarizing what was taught are absolutely necessary for English language learners.

General Information About the Student Populations (cont.)

Regular Education Students

By Wendy Conklin

Often, regular education students get overlooked when planning curriculum. More emphasis is usually placed on those who struggle and, at times, on those who excel. Teachers spend time teaching basic skills and even go below grade level to ensure that all students are up to speed. While this is a noble thing and is necessary at times, in the midst of it all, the regular education students can get lost in the shuffle. We must not forget that differentiated strategies are good for the on-grade level students, too. Providing activities that are too challenging can frustrate these students; on the other hand, assignments that are too easy can be boring and a waste of their time. The key to reaching this population successfully is to find just the right level of activities and questions while keeping a keen eye on their diverse learning styles.

Gifted Education Students

By Wendy Conklin

In recent years, many state and school district budgets have cut funding that has in the past provided resources for their gifted and talented programs. The push and focus of schools nationwide is proficiency. It is important that students have the basic skills to read fluently, solve math problems, and grasp science concepts. As a result, funding has been redistributed in hopes of improving test scores on state and national standardized tests. In many cases, the attention has focused only on improving low test scores to the detriment of the gifted students who need to be challenged.

Differentiating the products you require from your students is a very effective and fairly easy way to meet the needs of gifted students. Actually, this simple change to your assignments will benefit all levels of students in your classroom. While some students are strong verbally, others express themselves better through nonlinguistic representation. After reading the texts in this book, students can express their comprehension through different means, such as drawings, plays, songs, skits, or videos. It is important to identify and address different learning styles. By giving more open-ended assignments, you allow for more creativity and diversity in your classroom. These differentiated products can easily be aligned with content standards. To assess these standards, use differentiated rubrics.

Strategies for Using the Leveled Texts

Special Education Students

By Dennis Benjamin

Vocabulary Scavenger Hunt

A valuable prereading strategy is a Vocabulary Scavenger Hunt. Students preview the text and highlight unknown words. Students then write the words on specially divided pages. The pages are divided into quarters with the following headings: *Definition*, *Sentence*, *Examples*, and *Nonexamples*. A section called *Picture* is put over the middle of the chart.

Example Vocabulary Scavenger Hunt

astronomer

Definition	**Sentence**
a scientist who studies the universe and the objects within it	Astronomers use telescopes to discover new planets.
Examples	**Nonexamples**
Nicholas Copernicus; Galileo Galilei; Carl Sagan	George Washington; Ludwig van Beethoven; Rosa Parks

This encounter with new vocabulary enables students to use it properly. The definition identifies the word's meaning in student-friendly language. The sentence should be written so that the word is used in context. This helps the student make connections with background knowledge. Illustrating the sentence gives a visual clue. Examples help students prepare for factual questions from the teacher or on standardized assessments. Nonexamples help students prepare for ***not*** and ***except for*** test questions such as "All of these are explorers *except for*..." and "Which of these people is *not* an explorer?" Any information the student was unable to record before reading can be added after reading the text.

Strategies for Using the Leveled Texts (cont.)

Special Education Students (cont.)

Graphic Organizers to Find Similarities and Differences

Setting a purpose for reading content focuses the learner. One purpose for reading can be to identify similarities and differences. This is a skill that must be directly taught, modeled, and applied. The authors of *Classroom Instruction That Works* state that identifying similarities and differences "might be considered the core of all learning" (Marzano, Pickering, and Pollock 2001, 14). Higher-level tasks include comparing and classifying information and using metaphors and analogies. One way to scaffold these skills is through the use of graphic organizers, which help students focus on the essential information and organize their thoughts.

Example Classifying Graphic Organizer

Astronaut/ Cosmonaut	Nation	Major Space Achievement	Date of Achievement	Spacecraft/ Mission
Yuri Gagarin	Soviet Union	First person to travel in space	April 12, 1961	*Vostok 1*
Alan Shepard	United States	First American in space	May 5, 1961	*Freedom 7*
Alexei A. Leonov	Soviet Union	First spacewalk	March 18, 1965	*Voskhod 2*
Neil A. Armstrong	United States	First person on the moon	July 16, 1969	*Apollo 11*

The Riddles Graphic Organizer allows students to compare and contrast the astronauts using riddles. Students first complete a chart you've designed. Then, using that chart, they can write summary sentences. They do this by using the riddle clues and reading across the chart. Students can also read down the chart and write summary sentences. With the chart below, students could write the following sentences: Gagarin and Leonov represented the Soviet Union in space. Neil Armstrong and Alan Shepard walked on the moon.

Example Riddles Graphic Organizer

Who am I?	Gagarin	Shepard	Leonov	Armstrong
I walked on the moon.		x		x
I represented the Soviet Union in space.	x		x	
I was the first person from my nation to explore space.	x	x		
I was a pilot before I became an astronaut.	x	x	x	x
I went into space in 1961.	x	x		

Strategies for Using the Leveled Texts (cont.)

Special Education Students (cont.)

Framed Outline

This is an underused technique that bears great results. Many special education students have problems with reading comprehension. They need a framework to help them attack the text and gain confidence in comprehending the material. Once students gain confidence and learn how to locate factual information, the teacher can fade out this technique.

There are two steps to successfully using this technique. First, the teacher writes cloze sentences. Second, the students complete the cloze activity and write summary sentences.

Example Framed Outline

On July 21, 1969, the first _____ walked on the moon. His name was Neil _____ . He and astronaut Edwin E. "Buzz" Aldrin Jr. spent more than two hours _____ on the moon. They wore bulky _____ .

Summary Sentences:

On July 21, 1969, U.S. astronauts Neil Armstong and Edwin E. Adrin Jr. became the first humans to step foot on the moon. The Apollo 11 astronauts trained for years before becoming space pioneers.

Modeling Written Responses

A frequent criticism heard by special educators is that special education students write poor responses to content-area questions. This problem can be remedied if special education and classroom teachers model what good answers look like. While this may seem like common sense, few teachers take the time to do this. They just assume all children know how to respond in writing.

This is a technique you may want to use before asking your students to respond to the comprehension questions associated with the leveled texts in this series. First, read the question aloud. Then, write the question on an overhead and talk aloud about how you would go about answering the question. Next, write the answer using a complete sentence that accurately answers the question. Repeat the procedure for several questions so that students make the connection that quality written responses are your expectation.

Strategies for Using the Leveled Texts (cont.)

English Language Learners

By Marcela von Vacano

Effective teaching for English language learners (ELLs) requires effective planning. In order to achieve success, teachers need to understand and use a conceptual framework to help them plan lessons and units. There are six major components to any framework. Each is described in more detail below.

1. Select and Define Concepts and Language Objectives—Before having students read one of the texts in this book, the teacher must first choose a science concept and language objective (reading, writing, listening, or speaking) appropriate for the grade level. Then, the next step is to clearly define the concept to be taught. This requires knowledge of the subject matter, alignment with local and state objectives, and careful formulation of a statement that defines the concept. This concept represents the overarching idea. The science concept should be written on a piece of paper and posted in a visible place in the classroom.

By the definition of the concept, post a set of key language objectives. Based on the content and language objectives, select essential vocabulary from the text. The number of new words selected should be based on students' English language levels. Post these words on a word wall that may be arranged alphabetically or by themes.

2. Build Background Knowledge—Some ELLs may have a lot of knowledge in their native language, while others may have little or no knowledge. The teacher will want to build the background knowledge of the students using different strategies such as the following:

Visuals: Use posters, photographs, postcards, newspapers, magazines, drawings, and video clips of the topic you are presenting. The texts in this series include multiple primary sources for your use.

Realia: Bring real-life objects to the classroom. If you are teaching about the plant life cycle, bring in items such as soil, seeds, roots, leaves, and flowers.

Vocabulary and Word Wall: Introduce key vocabulary in context. Create families of words. Have students draw pictures that illustrate the words and write sentences about the words. Also be sure you have posted the words on a word wall in your classroom.

Desk Dictionaries: Have students create their own desk dictionaries using index cards. On one side, they should draw a picture of the word. On the opposite side, they should write the word in their own language and in English.

Strategies for Using the Leveled Texts (cont.)

English Language Learners (cont.)

3. Teach Concepts and Language Objectives—The teacher must present content and language objectives clearly. He or she must engage students using a hook and must pace the delivery of instruction, taking into consideration students' English language levels. The concept or concepts to be taught must be stated clearly. Use the first languages of the students whenever possible or assign other students who speak the same languages to mentor and to work cooperatively with the ELLs.

Lev Semenovich Vygotsky, a Russian psychologist, wrote about the Zone of Proximal Development (ZPD). This theory states that good instruction must fill the gap that exists between the present knowledge of a child and the child's potential. Scaffolding instruction is an important component when planning and teaching lessons. ELLs cannot jump stages of language and content development. You must determine where the students are in the learning process and teach to the next level using several small steps to get to the desired outcome. With the leveled texts in this series and periodic assessment of students' language levels, teachers can support students as they climb the academic ladder.

4. Practice Concepts and Language Objectives—ELLs need to practice what they learn with engaging activities. Most people retain knowledge best after applying what they learn to their own lives. This is definitely true for English language learners. Students can apply content and language knowledge by creating projects, stories, skits, poems, or artifacts that show what they learned. Some activities should be geared to the right side of the brain, like those listed above. For students who are left-brain dominant, activities such as defining words and concepts, using graphic organizers, and explaining procedures should be developed. The following teaching strategies are effective in helping students practice both language and content:

> **Simulations**: Students learn by doing. For example, when teaching about the plant life cycle, you can have students figure out what they would need to grow a plant. First, they need to make a list and collect the necessary items, such as a clay pot or empty milk jug, planting soil, seeds, and water. They can fill the pot or jug with soil and plant a seed. They will need to water the plant daily, and make sure it gets enough sun. Lastly, students can measure and record how much the plant grows in a week, two weeks, or one month.

> **Literature response**: Read a text from this book. Have students choose two people described or introduced in the text. Ask students to create a conversation the people might have. Or, you can have students write journal entries about events in the daily lives of the famous scientists.

Strategies for Using the Leveled Texts

English Language Learners (cont.)

4. Practice Concepts and Language Objectives (cont.)

Have a short debate: Make a controversial statement such as, "It isn't necessary for humans to explore space." After reading a text in this book, have students think about the question and take a position. As students present their ideas, one student can act as a moderator.

Interview: Students may interview a member of the family or a neighbor in order to obtain information regarding a topic from the texts in this book. For example: What was the reaction when Apollo 11 astronauts walked on the moon?

5. Evaluation and Alternative Assessments

We know that evaluation is used to inform instruction. Students must have the opportunity to show their understanding of concepts in different ways and not only through standard assessments. Use both formative and summative assessment to ensure that you are effectively meeting your content and language objectives. Formative assessment is used to plan effective lessons for a particular group of students. Summative assessment is used to find out how much the students have learned. Other authentic assessments that show day-to-day progress are: text retelling, teacher rating scales, student self-evaluations, cloze testing, holistic scoring of writing samples, performance assessments, and portfolios. Periodically assessing student learning will help you ensure that students continue to receive the correct levels of texts.

6. Home-School Connection

The home-school connection is an important component in the learning process for ELLs. Parents are the first teachers, and they establish expectations for their children. These expectations help shape the behavior of their children. By asking parents to be active participants in the education of their children, students get a double dose of support and encouragement. As a result, families become partners in the education of their children and chances for success in your classroom increase.

You can send home copies of the texts in this series for parents to read with their children. You can even send multiple levels to meet the needs of your second-language parents as well as your students. In this way, you are sharing your science content standards with your whole second-language community.

Strategies for Using Leveled Texts (cont.)

provide a ceiling that is too low for gifted students become bored. We know more? Offering open-ended questions will give high-ability students the opportunities to perform at or above their ability levels. For example, ask students to evaluate scientific topics described in the texts, such as: "Do you think the United States should be continuing space exploration?" or "What do you think our government should do to deal with global warming?" These questions require students to form opinions, think deeply about the issues, and form pro and con statements in their minds. To questions like these, there really is not one right answer.

The generic, open-ended question stems listed below can be adapted to any topic. There is one leveled comprehension question for each text in this book. These question stems can be used to develop further comprehension questions for the leveled texts.

- In what ways did…
- How might you have done this differently…
- What if…
- What are some possible explanations for…
- How does this affect…
- Explain several reasons why…
- What problems does this create…
- Describe the ways…
- What is the best…
- What is the worst…
- What is the likelihood…
- Predict the outcome…
- Form a hypothesis…
- What are three ways to classify…
- Support your reason…
- Compare this to modern times…
- Make a plan for…
- Propose a solution…
- What is an alternative to…

14

#50160—Leveled Texts for Science: Earth & Space Science © Shell Education

Strategies for Using the Leveled Texts (cont.)

Gifted Education Students (cont.)

Student-Directed Learning

Because they are academically advanced, gifted students are often the leaders in classrooms. They are more self-sufficient learners, too. As a result, there are some student-directed strategies that teachers can employ successfully with these students. Remember to use the texts in this book as jumpstarts so that students will be interested in finding out more about the science concepts presented. Gifted students may enjoy any of the following activities:

- Writing their own questions, exchanging their questions with others, and grading the responses.
- Reviewing the lesson and teaching the topic to another group of students.
- Reading other nonfiction texts about these science concepts to further expand their knowledge.
- Writing the quizzes and tests to go along with the texts.
- Creating illustrated time lines to be displayed as visuals for the entire class.
- Putting together multimedia presentations about the scientific breakthroughs and concepts.

Tiered Assignments

Teachers can differentiate lessons by using tiered assignments, or scaffolded lessons. Tiered assignments are parallel tasks designed to have varied levels of depth, complexity, and abstractness. All students work toward one goal, concept, or outcome, but the lesson is tiered to allow for different levels of readiness and performance levels. As students work, they build on their prior knowledge and understanding. Students are motivated to be successful according to their own readiness and learning preferences.

Guidelines for writing tiered lessons include the following:

1. Pick the skill, concept, or generalization that needs to be learned.

2. Think of an on-grade-level activity that teaches this skill, concept, or generalization.

3. Assess the students using classroom discussions, quizzes, tests, or journal entries and place them in groups.

4. Take another look at the activity from Step 2. Modify this activity to meet the needs of the below-grade-level and above-grade-level learners in the class. Add complexity and depth for the gifted students. Add vocabulary support and concrete examples for the below-grade-level students.

How to Use This Product

Readability Chart

Title of the Text	Star	Circle	Square	Triangle
Jet Streams and Trade Winds	1.9	3.1	4.7	6.5
The Water Cycle	1.6	3.1	4.7	6.5
Tornadoes and Hurricanes	1.8	3.4	4.6	6.7
Structure of the Earth	2.1	3.5	5.2	6.5
Earthquakes and Volcanoes	2.2	3.0	4.9	6.6
Plate Tectonics	2.2	3.5	5.1	7.0
Wegener Solves a Puzzle	2.0	3.5	5.1	6.6
The Rock Cycle	1.8	3.4	4.5	7.1
Fun with Fossils	2.0	3.2	5.0	6.9
The Inner Planets	2.2	3.1	4.8	6.7
The Outer Planets	2.1	3.4	4.9	6.5
Our Place in Space	2.2	3.4	5.2	6.7
Other Citizens of the Solar System	2.1	3.1	5.0	6.9
The Astronomer's Toolbox	2.2	3.4	4.8	6.8
The Journey to Space	1.7	3.1	4.9	6.7

Components of the Product

Primary Sources

- Each level of text includes multiple primary sources. These documents, photographs, and illustrations add interest to the texts. The scientific images also serve as visual support for second language learners. They make the texts more context rich and bring the texts to life.

How to Use This Product (cont.)

Components of the Product (cont.)

Comprehension Questions

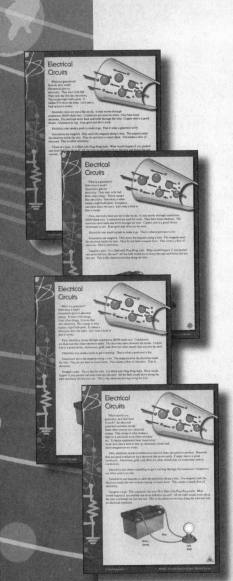

- Each level of text includes one comprehension question. Like the texts, the comprehension questions were leveled by an expert. They are written to allow all students to be successful within a whole-class discussion. The questions for the same topic are closely linked so that the teacher can ask a question on that topic and all students will be able to answer. The lowest-level students might focus on the facts, while the upper-level students can delve deeper into the meanings.

- Teachers may want to base their whole-class question on the square level questions. Those were the starting points for all the other leveled questions.

The Levels

- There are 15 topics in this book. Each topic is leveled to four different reading levels. The images and fonts used for each level within a topic look the same.

- Behind each page number, you'll see a shape. These shapes indicate the reading levels of each piece so that you can make sure students are working with the correct texts. The reading levels fall into the ranges indicated to the left. See the chart at left for specific levels of each.

Leveling Process

- The texts in this series are taken from the Primary Source Readers kits published by Teacher Created Materials. A reading expert went through the texts and leveled each one to create four distinct reading levels.

- After that, a special education expert and an English language learner expert carefully reviewed the lowest two levels and suggested changes that would help their students comprehend the texts better.

- The texts were then leveled one final time to ensure the editorial changes made during the process kept them within the ranges described to the left.

Levels 1.5–2.2

Levels 3.0–3.5

Levels 4.5–5.2

Levels 6.5–7.2

How to Use This Product (cont.)

Tips for Managing the Product

How to Prepare the Texts

- When you copy these texts, be sure you set your copier to copy photographs. Run a few test pages and adjust the contrast as necessary. If you want the students to be able to appreciate the images, you need to carefully prepare the texts for them.

- You also have full-color versions of the texts provided in PDF form on the CD. (See page 144 for more information.) Depending on how many copies you need to make, printing the full-color versions and copying those might work best for you.

- Keep in mind that you should copy two-sided to two-sided if you pull the pages out of the book. The shapes behind the page numbers will help you keep the pages organized as you prepare them.

Distributing the Texts

Some teachers wonder about how to hand the texts out within one classroom. They worry that students will feel insulted if they do not get the same papers as their neighbors. The first step in dealing with these texts is to set up your classroom as a place where all students learn at their individual instructional levels. Making this clear as a fact of life in your classroom is key. Otherwise, the students may constantly ask about why their work is different. You do not need to get into the technicalities of the reading levels. Just state it as a fact that every student will not be working on the same assignment every day. If you do this, then passing out the varied levels is not a problem. Just pass them to the correct students as you circle the room.

If you would rather not have students openly aware of the differences in the texts, you can try these ways to pass out the materials:

- Make a pile in your hands from star to triangle. Put your finger between the circle and square levels. As you approach each student, you pull from the top (star), above your finger (circle), below your finger (square), or the bottom (triangle). If you do not hesitate too much in front of each desk, the students will probably not notice.

- Begin the class period with an opening activity. Put the texts in different places around the room. As students work quietly, circulate and direct students to the right locations for retrieving the texts you want them to use.

- Organize the texts in small piles by seating arrangement so that when you arrive at a group of desks you have just the levels you need.

How to Use This Product (cont.)

Correlation to Standards

The No Child Left Behind (NCLB) legislation mandates that all states adopt academic standards that identify the skills students will learn in kindergarten through twelfth grade. While many states had already adopted academic standards prior to NCLB, the legislation set requirements to ensure the standards were detailed and comprehensive.

Standards are designed to focus instruction and guide adoption of curricula. Standards are statements that describe the criteria necessary for students to meet specific academic goals. They define the knowledge, skills, and content students should acquire at each level. Standards are also used to develop standardized tests to evaluate students' academic progress.

In many states today, teachers are required to demonstrate how their lessons meet state standards. State standards are used in the development of Shell Education products, so educators can be assured that they meet the academic requirements of each state.

How to Find Your State Correlations

Shell Education is committed to producing educational materials that are research and standards based. In this effort, all products are correlated to the academic standards of the 50 states, the District of Columbia, and the Department of Defense Dependent Schools. A correlation report customized for your state can be printed directly from the following website: **http://www.shelleducation.com**. If you require assistance in printing correlation reports, please contact Customer Service at 1-800-877-3450.

McREL Compendium

Shell Education uses the Mid-continent Research for Education and Learning (McREL) Compendium to create standards correlations. Each year, McREL analyzes state standards and revises the compendium. By following this procedure, they are able to produce a general compilation of national standards.

Each reading comprehension strategy assessed in this book is based on one or more McREL content standards. The following chart shows the McREL standards that correlate to each lesson used in the book. To see a state-specific correlation, visit the Shell Education website at **http://www.shelleducation.com**.

McREL	Benchmark	Text
1.1	Knows the composition and structure of the Earth's atmosphere	Jet Streams and Trade Winds
1.2	Knows the processes involved in the water cycle and their effects on climatic patterns	The Water Cycle
1.3	Knows that the Sun is the principle energy source for phenomena on the Earth's surface	The Water Cycle
1.4	Knows factors that can impact the Earth's climate	Tornadoes and Hurricanes
1.5	Knows how the tilt of the Earth's axis and the Earth's revolution around the Sun affect seasons and weather patterns	Our Place in Space
1.6	Knows ways in which clouds affect weather and climate	Tornadoes and Hurricanes
1.7	Knows the properties that make water an essential component of the Earth system	The Water Cycle
2.1	Knows that the Earth is comprised of layers including a core, mantle, lithosphere, hydrosphere, and atmosphere	Structure of the Earth
2.2	Knows how land forms are created through a combination of constructive and destructive forces	Plate Tectonics
2.3	Knows components of soil and other factors that influence soil texture, fertility, and resistance to erosion	The Rock Cycle
2.4	Knows that the Earth's crust is divided into plates that move at extremely slow rates in response to movements in the mantle	Plate Tectonics
2.5	Knows processes involved in the rock cycle	The Rock Cycle
2.6	Knows that sedimentary, igneous, and metamorphic rocks contain evidence of the minerals, temperatures, and forces that created them	The Rock Cycle
2.7	Knows how successive layers of sedimentary rock and the fossils contained within them can be used to confirm the age, history, and changing life forms of the Earth, and how this evidence is affected by the folding, breaking, and uplifting of layers	Fun with Fossils
2.8	Knows that fossils provide important evidence of how environmental conditions have changed on the Earth over time	Fun with Fossils
3.1	Knows characteristics and movement patterns of the nine planets in our solar system	The Inner Planets; The Outer Planets
3.2	Knows how the regular and predictable motions of the Earth and moon explain phenomena on Earth	Our Place in Space
3.3	Knows characteristics of the sun and its position in the universe	Our Place in Space
3.4	Knows that gravitational force keeps planets in orbit around the Sun and moons in orbit around the planets	The Inner Planets; The Outer Planets; Other Citizens of the Solar System
3.5	Knows characteristics and movement patterns of asteroids, comets, and meteors	Other Citizens of the Solar System
3.6	Knows that the universe consists of many billions of galaxies and that incomprehensible distances separate these galaxies and stars from one another and from the Earth	The Astronomer's Toolbox
3.7	Knows that the planet Earth and our solar system appear to be somewhat unique, although similar systems might yet be discovered in the universe	The Inner Planets; Our Place in Space

Jet Streams and Trade Winds

Some days there are clouds in the sky. Some days there are no clouds. Have you ever wondered why? Where do the clouds come from? Where do they go? You might know the answer. Winds move clouds across the sky. You might not know why it happens. All weather takes place in the layer of air closest to Earth's surface. Lots of things change weather such as heat, water, and wind.

Jet Streams

The sun warms Earth's surface. This makes heat and moisture rise into the sky. There are four main jet streams there, high in the sky. Jet streams are rivers of wind. They are thousands of miles long, hundreds of miles wide, and several miles deep.

Jet streams are one of the things that change our weather. They lower the heat. They move the moisture around. They blow about 200 kilometers per hour (125 miles per hour). They bend and move. They don't stay in the same spot. They move toward the equator. Then they move away from it.

Jet streams weren't known about until modern times. When jet planes were invented, they flew high enough to find the jet streams. In fact, that is where they got their names.

Trade Winds

Before scientists knew about jet streams, they studied trade winds and monsoons. Trade winds are winds that blow east to west. They blow the same way all year long. Monsoons are giant rain storms.

A scientist named Edmond Halley studied these parts of the weather. In 1686, he wrote a paper. He had a chart on the trade winds and monsoons. He wrote that the sun creates most weather. He was right. He showed how the weather came from air pressure. It also comes from altitude. Altitude is how high a place is above sea level.

Halley's work on trade winds was not complete. It did not explain the winds' pattern. They always blow from east to west. George Hadley lived about the same time. He was a lawyer and a scientist. He explained the pattern of trade winds.

The sun evaporates a lot of water near the equator. The water vapor is warm. It rises into the sky. It flows north and south. This water goes very far. Then it cools. It falls back to Earth. The falling rain pushes air down. Wind is caused by the air falling. The way Earth rotates on its axis causes the wind to blow east and west.

Comprehension Question

What makes wind?

Jet Streams and Trade Winds

Some days there are clouds in the sky. Some days there are no clouds. Have you ever wondered why? Where do the clouds come from? Where do they go? You probably know the answer. Winds move clouds across the sky. What you may not know is why it happens. All weather takes place in the layer of the atmosphere closest to Earth's surface. Lots of things change weather. The biggest factors are heat, water, and wind.

Jet Streams

The sun warms Earth's surface, making heat and moisture rise into the atmosphere. There are four main jet streams there, high in the sky. Jet streams are rivers of wind. They are thousands of miles long, hundreds of miles wide, and several miles deep.

Jet streams are one of the driving forces of weather. The jet streams lower the heat and move the moisture around. They blow about 200 kilometers per hour (125 miles per hour). They bend and move. They don't stay in the same spot. They move toward the equator or away from it.

Jet streams weren't discovered until modern times. Jet planes were invented and flew high enough to find the jet streams. In fact, that is where they got their names.

Trade Winds

Before scientists knew about jet streams, they studied trade winds and monsoons. Trade winds are winds that blow east to west. They blow in regular patterns. Monsoons are big, violent rainstorms.

A scientist named Edmond Halley studied these parts of the weather. In 1686, he wrote a paper. In it, he had a chart on the trade winds and monsoons. Halley wrote that the sun creates most of the weather on Earth. He was right. He also showed how air pressure and altitude affect the weather. Altitude is a place's height above sea level.

Halley's theory on Earth's trade winds was not complete. It did not explain their pattern. The trade winds always blow from east to west. George Hadley lived about the same time as Halley. He was a lawyer and a scientist. He explained trade winds even better.

He explained that the sun evaporates lots of water near the equator. The water vapor in the warmer air would rise up into the sky and flow north and south. This water travels a long distance. When it cools, it falls back to Earth. The falling rain pushes air down. Wind is caused by the air falling. The way Earth rotates on its axis causes the wind to blow east and west.

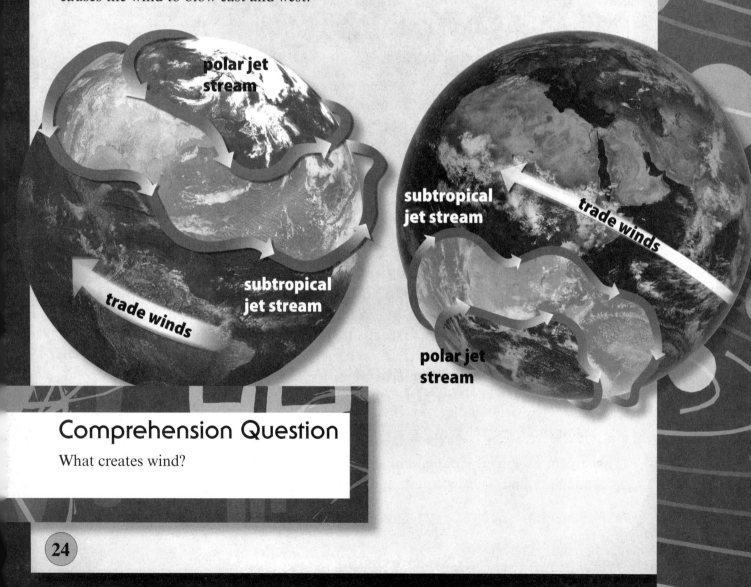

Comprehension Question

What creates wind?

Jet Streams and Trade Winds

Have you ever wondered why some days there are clouds in the sky and some days there aren't? Where do the clouds come from, and where do they go to? You probably know the answer: winds move clouds across the planet. What you may not know is that all weather happens in the layer of atmosphere closest to Earth's surface. Many things affect weather. The biggest factors are heat, water, and wind.

Jet Streams

The sun warms Earth's surface. This makes heat and water rise into the atmosphere. There are four main jet streams there, high in the sky. Jet streams are rivers of wind. They are thousands of miles long, hundreds of miles wide, and several miles deep.

Jet streams are one of the driving forces of weather changes. They lower the heat and move around the moisture. They blow an average of 200 kilometers per hour (125 miles per hour). They bend and move in different ways. They don't always stay in the same spot. They move either toward the equator or away from it.

Jet streams weren't discovered until the modern era, when jet planes were invented and flew high enough to find them. In fact, that is where they got their names.

Trade Winds

Before scientists knew about jet streams, they tried to understand trade winds and monsoons. Trade winds are winds that blow mainly east to west in regular patterns. Monsoons are big, violent rainstorms.

Halley's work in water science relates mainly to weather. In 1686, Edmond Halley published an important paper and a chart on the trade winds and monsoons. In this same paper, Halley wrote that the sun is the driving force behind most of the weather on Earth. He was right. He also showed the relationship between air pressure, altitude (height above sea level), and weather. Air pressure and altitude affect the weather.

Halley's theory on Earth's trade winds couldn't explain why they always blew from east to west. George Hadley lived about the same time as Halley. He was a lawyer and amateur scientist. He came up with a more complete explanation of why trade winds happen.

He explained that the sun would evaporate a great deal of water near the equator. The water vapor in the warmer air would rise up into the atmosphere and flow north and south. This water travels a long distance in the atmosphere and cools. Because it cools, it falls back to Earth. The falling rain pushes air down. Wind is caused by this motion of the air falling. The way Earth rotates on its axis causes the wind to blow east and west.

Comprehension Question

Where does the energy in the wind come from?

Jet Streams and Trade Winds

Have you ever wondered why some days there are clouds in the sky and some days there aren't? You probably know why: winds move clouds across the planet. What you may not know is that all weather happens in the layer of atmosphere closest to Earth's surface. Many things affect weather. The biggest factors are heat, water, and wind.

Jet Streams

The sun warms the Earth's surface, which evaporates water into vapor. The vapor rises high into the sky, where four rivers of wind called jet streams flow. Thousands of miles long, hundreds of miles wide, and several miles deep, jet streams are a driving force of weather changes. The jet streams lower the vapor's heat and move it across the planet. Jet streams blow 200 kilometers per hour (125 miles per hour) on average. Jet streams always flow parallel to the equator, but they move north and south over the course of the year.

Jet streams weren't discovered until the modern era, when newly invented jet planes flew high enough to find them. In fact, that is where they got their names.

Trade Winds

Before scientists knew about jet streams, they tried to understand trade winds and monsoons. Trade winds blow primarily east to west in regular patterns; monsoons are big, violent rainstorms that occur seasonally.

Halley's work in water science relates mainly to weather. In 1686, Edmond Halley published an important paper and a chart on the trade winds and monsoons. In this same paper, Halley wrote that the sun is the driving force behind most of the weather on Earth. He was right. He also showed the relationship between air pressure, altitude (height above sea level), and weather. Air pressure and altitude affect the weather.

Halley's theory on Earth's trade winds couldn't explain why they always blew in the same direction, east to west. George Hadley was a lawyer and amateur scientist who lived about the same time as Halley. He figured out the cause behind trade winds' constant direction.

Hadley explained that the sun evaporates a great deal of water near the equator. The water vapor in the warmer air would rise up into the atmosphere and flow north and south away from the equator. Eventually, this water cools and falls back to Earth. The falling rain pushes air down, and wind is caused by this motion of the air falling. Since the Earth rotates on its axis, it "spins" that wind, causing it to blow east and west.

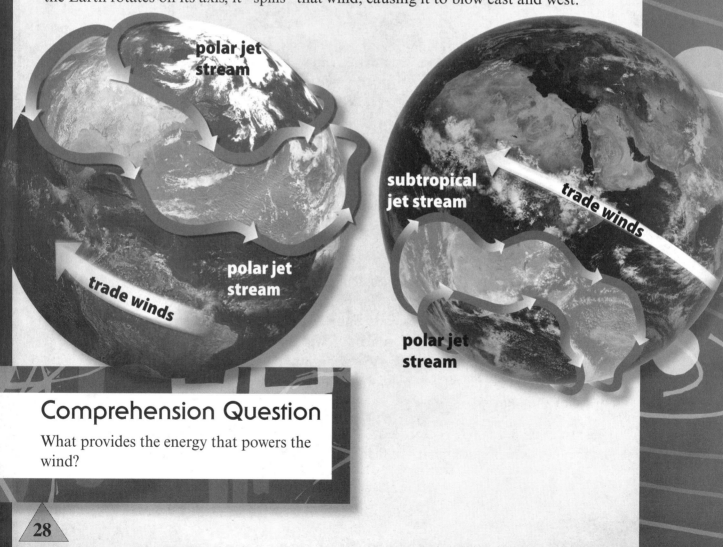

Comprehension Question

What provides the energy that powers the wind?

The Water Cycle

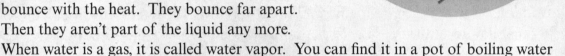

The water cycle is a circle. There is no real start to it. We must pick a place to start. Evaporation is as good a place as any to start.

Evaporation

Evaporation is all around us. You can heat a liquid. Then it will change to a gas. The warmed molecules in the liquid move. They bounce with the heat. They bounce far apart. Then they aren't part of the liquid any more. When water is a gas, it is called water vapor. You can find it in a pot of boiling water on the stove. You can find it when the sun heats water in the oceans.

When the water vapor moves up into the air, it loses its heat. Then it turns back to a liquid. The water clumps up. It forms small drops or ice crystals. They are not big enough to fall back to Earth. When there are enough of them close together, they can form clouds.

The sun heats Earth. It doesn't heat evenly. Some places get hot. Some places don't. The hot air puts pressure on the cold air. The pressure must balance out. Air moves from high-pressure parts to low-pressure parts. This makes wind. Earth's spin twists the air. Currents in the oceans move the air, too. Air moving from side to side is called advection. It is why clouds move.

Precipitation

Water vapor in the sky can form water droplets. It can also form ice crystals. Wind causes the water drops and ice to bump into each other. They form larger clumps. The large clumps make even larger clumps. They fall to the earth. This is called precipitation. We know it as rain, snow, sleet, and hail.

Now the water is on the ground. This is the next phase of the water cycle. What happens depends on where it falls and in what form. The water could be ice as snow, sleet, or hail. It might pile up and stay ice for a while. It may melt and change to liquid water. The water could fall as rain. It can soak into the ground or it can run off and form streams or rivers.

Groundwater

When water soaks into the ground, it flows into tiny spaces. The spaces are in between bits of soil. Deeper down the water can't flow through rock. The rock is impermeable. The water is trapped. It fills up all the spaces in the soil above. That rock is permeable. Then it is called groundwater. The water backs up. It spills out. It starts moving downhill.

Sometimes there is a lot of rain from a storm. The water can't all soak into the ground. The water runs over the ground's surface. It runs into streams. It runs into rivers. The water keeps going and going. Over time, all the water makes its way back to the ocean.

Water flows to the oceans. It has gone all the way around the water cycle. All the water that flows into the ocean once came out of it.

Comprehension Question

Write three things that happen to a drop of water in the water cycle.

The Water Cycle

There is really no start to the water cycle. To talk about it, we must pick a place to start. Evaporation is as good a place as any.

Evaporation

Evaporation happens all over. When a liquid is heated, it changes to a gas. The heated molecules move around very quickly. They move far apart. Then they aren't part of the liquid any more. When water is a gas, it is called water vapor. It is made in a pot of boiling water on the stove. It is made when the sun heats water in the oceans.

The water vapor moves up into the air. It loses the heat it had taken in. Then the vapor turns back into a liquid. The water starts to clump up. It forms small droplets or ice crystals. The droplets or crystals are very tiny. They are not heavy enough to fall back to Earth. When there are enough of them close together, they can form clouds.

When the sun heats Earth, it doesn't heat evenly. Some places get hot. Some places don't. The hot air puts pressure on the cold air. To make the pressure balance, air moves from areas with high pressure to areas with low pressure, creating wind. The spinning of Earth and currents in the oceans can affect movement of the air, too. Air moving from side to side is called advection. It is why clouds move across the planet.

Precipitation

Water vapor in the sky can form water droplets. It can also turn into solid ice crystals. Wind and air movement cause the drops and ice to bump into each other. They form larger clumps. If they get large enough, they fall to the earth. This is called precipitation. Of course, it is better known as rain, snow, sleet, and hail.

Water on the ground is the next phase of the water cycle. What happens depends on where it falls and in what form. The water could be ice as snow, sleet, or hail. It might pile up and stay ice for a while. It may melt and change to liquid water. The water could fall as rain. It can soak into the ground or it can run off and form streams or rivers.

Groundwater

When water soaks into the ground, it flows into tiny spaces. The spaces are in between bits of soil. Even deeper down, the rock is impermeable. That means the water can't flow through it. It is trapped. That water permeates the soil above. That means it fills up all the spaces in the soil. Water that soaks into the ground is called groundwater. The water overflows. It starts moving downhill.

Sometimes there is a lot of water during a rainstorm. The water can't all soak into the ground. Instead, the rainwater runs over the ground's surface. It runs into streams. It runs into rivers. With enough time, all the water makes its way back to the ocean.

Water flows to the oceans. It has gone all the way around the water cycle. All the water that flows into the ocean once came out of the ocean.

Comprehension Question

Describe three things that happen to a drop of water in the water cycle.

The Water Cycle

There's really no start to the water cycle. To talk about it, we must start somewhere. Evaporation is as good a place as any.

Evaporation

Evaporation happens all over. When a liquid is heated enough, it changes to a gas. The heated molecules move around very fast. They move too far apart to be a part of the liquid. When water evaporates, we call it water vapor. It happens on a small scale when a stove heats a pot of water. It happens on a very large scale when the sun heats water in the oceans.

The water vapor moves up through the atmosphere. It loses the heat it had taken in. When it loses enough heat, the vapor turns back into a liquid. The water molecules start sticking together. They form small droplets or ice crystals. The droplets or crystals are very tiny and not heavy enough to fall back to Earth. When there are enough of them close together, they can form clouds.

When the sun heats Earth, it doesn't heat evenly. Some places get hotter than other places do. This causes pressure differences in the air. To make the pressure balance, air moves from areas with high pressure to areas with low pressure, creating wind. The spinning of Earth and currents in the oceans can affect movement of the air, too. This process of air moving from side to side across the earth is called advection. It is why clouds move across the planet.

Precipitation

Water vapor in the sky can form water droplets. It can also turn into solid ice crystals. Wind and air movement cause these particles to bump into each other. They form larger particles. If they get large enough, they fall to the earth as precipitation. Of course, this is better known as rain, snow, sleet, and hail.

Water progresses to the next phase of the water cycle once it hits the ground. What happens depends on where it falls and in what form. If the water is frozen as snow, sleet, or hail, it might pile up and stay frozen for a while. It may melt quickly and change to liquid water. When water falls as rain, it can soak into the ground or it can run off and form streams or rivers.

Groundwater

When water soaks into the ground, it flows into tiny spaces between soil particles. Deeper underground, the rock is impermeable. The water can't flow through it and is trapped. The captured water permeates the soil above. It fills up all the spaces between soil particles. Water that soaks into the ground like this is called groundwater. The water overflows. It starts moving horizontally.

If there is a lot of water during a rainstorm, the water can't all soak into the ground. Instead, the rainwater runs over the ground's surface. It collects into streams and rivers. Eventually, all the water makes its way back to the ocean.

Water flows to the oceans. It has traveled all the way around the water cycle. All the water that flows into the ocean once came out of the ocean.

Comprehension Question

Describe the trip that one drop of water makes as it goes around the water cycle.

The Water Cycle

There's really no start to the water cycle, but to understand it, we must begin somewhere. Evaporation is as good a place as any.

Evaporation

Evaporation happens everywhere. When a liquid is heated enough, it changes to a gas. This happens when the heated molecules move around so fast they are no longer close enough together to be a part of the liquid. When water evaporates, we call it water vapor. It happens on a small scale when a stove heats a pot of water. It happens on a very large scale when the sun heats water in the oceans.

The water vapor moves up through the atmosphere and loses the heat it had taken in. When it loses enough heat, the vapor condenses back into a liquid. The water molecules start sticking together, and they form small droplets or ice crystals. The droplets or crystals are very tiny and not heavy enough to fall back to Earth. When there are enough of them close together, they can form clouds.

When the sun heats Earth, it doesn't heat evenly. Some places get hotter than other places do, and this causes pressure differences in the air. To make up for these differences, air moves from areas with high pressure to areas with low pressure, creating wind. In addition, the spinning of Earth and currents in the oceans can affect movement of the air on Earth as well. This process of air moving from side to side across the earth is called advection, and it is why clouds move across the planet.

Precipitation

Water vapor in the atmosphere can form water droplets or turn into solid ice crystals. Wind and air movement causes these particles to bump into each other, forming larger particles. If they get large enough, they fall to the earth as precipitation. Of course, precipitation is better known as rain, snow, sleet, and hail.

Water progresses to the next phase of the water cycle once it hits the ground. What happens depends on where it falls and in what form. If the water is frozen as snow, sleet, or hail, it might pile up and stay frozen for a while. It may melt quickly and change to liquid water. When water falls as rain, it can soak into the ground or it can run off and form streams or rivers.

Groundwater

When water soaks into the ground, it flows into tiny spaces. The spaces are in between bits of soil. Deeper down the water can't flow through rock. The rock is impermeable. The water is trapped. It fills up all the spaces in the soil above. That rock is permeable. Then it is called groundwater. The water backs up. It spills out. It starts moving downhill.

Sometimes there is a lot of rain from a storm. The water can't all soak into the ground. The water runs over the ground's surface. It runs into streams. It runs into rivers. The water keeps going and going. Over time, all the water makes its way back to the ocean.

Water flows to the oceans. It has gone all the way around the water cycle. All the water that flows into the ocean once came out of it.

Comprehension Question

Describe the water cycle from the point of view of a single drop of water.

Tornadoes and Hurricanes

A tornado is a huge storm. It acts like a huge vacuum. It moves at high speed. It goes as fast as fifty kilometers (thirty miles) per hour. It smashes things in its path. It lifts cars into the air and drops them far away.

Tornadoes can be found across the world. Many hit the central states in the United States. They have the most tornadoes on Earth. There are about a thousand tornadoes each year. Most are weak. Others are strong. They can do a great deal of damage.

Tornadoes start as thunderstorms. These huge storms can form supercells. That is when the storm starts to turn. It can be very dangerous. It can cause hail and strong winds. It can cause tornadoes.

Strong winds blow around the storm. Some strong winds go one way. Other strong winds go the other way. The air inside the clouds starts to spin. The spinning winds can touch the ground. That makes a tornado.

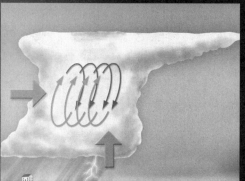

A thunderstorm can develop a supercell inside it. Winds go up, over, down, and back in a circle.

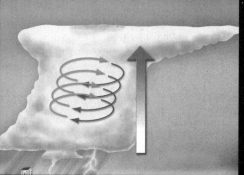

The supercell can get tilted up and down by a strong updraft.

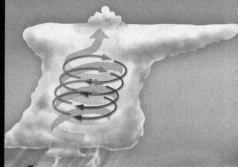

If the vertical supercell keeps going, it starts sucking wind and everything else up the cell. A tornado is born.

Hurricanes

As bad as tornadoes are, they aren't the worst storms. Each year hurricanes cause more damage than all other storms combined. Why? Hurricanes are huge, and they last a long time. Hurricanes start as tropical storms. They form in late summer or fall over warm water. The center of a hurricane is called the eye. It has very low pressure. Clouds rush toward it. But they start to spin due to Earth's rotation. As a result, the eye stays calm. It has no clouds and no wind. When people talk about the calm in the eye of the storm, that's what they mean.

Hurricanes also strike Australia. There they are called willy-willies. When they hit Asia, they are called typhoons. Cyclones form in the Indian Ocean. What they are called may change. They all act the same. These storms pull water from the sea. Then they dump it on land. The water they drop isn't salty. It is not like sea water. The ocean water evaporates. It leaves its salt behind.

Then the rain stops. The sky gets calm. The sun shines. You may think it's all over. But this is just the eye of the storm. The other side of the storm is about to hit! Hurricanes slowly die as they move inland. They leave a lot of damage.

High winds are not the only problem. The storm pushes water in front of it. This water comes close to land. It piles up in a wave. It is called a storm surge. This causes huge flooding. The water level rises. Small buildings near the shore are suddenly underwater. Then, the water goes back to the ocean. It pulls things like cars and homes out to sea.

Scientists give names to hurricanes. That is how they keep track of them. For the Atlantic Ocean there are six lists. Each has 21 names. They switch to a new list each year. Boy and girl names go from A to W. They skip Q and U. If all 21 names are used, the Greek letters name the rest. This happened for the first time in 2005. There were a record number of hurricanes.

So, what sorts of names are picked? Here is the list for the year 2005: Arlene, Bret, Cindy, Dennis, Emily, Franklin, Gert, Harvey, Irene, Jose, Katrina, Lee, Maria, Nate, Ophelia, Philippe, Rita, Stan, Tammy, Vince, Wilma, Alpha, Beta, Gamma, Delta, Epsilon, and Zeta. Was your name a hurricane that year?

Comprehension Question

How are tornadoes and hurricanes the same?

Tornadoes and Hurricanes

A tornado acts like a huge vacuum. It roars along at high speed. It goes as fast as fifty kilometers (thirty miles) per hour. It smashes everything in its path. It lifts cars into the air. Then it drops them far away.

Tornadoes can be found across the world. Many hit the central states in the United States. They have the most tornadoes on Earth. There are about a thousand tornadoes each year. Most are weak. Others are strong. They can do a great deal of damage.

Tornadoes start as thunderstorms. These huge storms can form supercells. That is when the storm starts to turn. It can be very dangerous. It can cause hail and strong winds. It can cause tornadoes.

Strong winds blow opposite each other. The air inside the clouds starts to spin. If it touches the ground, it makes a tornado.

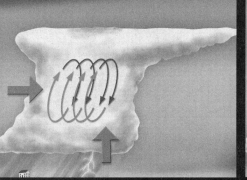

A thunderstorm can develop a supercell inside it. Winds go up, over, down, and back in a circle.

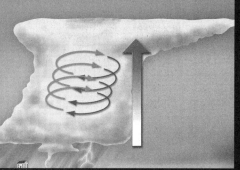

The supercell can get tilted up and down by a strong updraft.

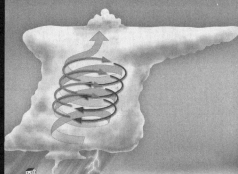

If the vertical supercell keeps going, it starts sucking wind and everything else up the cell. A tornado is born.

Hurricanes

As bad as tornadoes are, they aren't the worst storms. Each year hurricanes cause more damage than all other storms combined. Why? Hurricanes are huge, and they last a long time. Hurricanes start as tropical storms. They form in late summer or fall over warm water. The center of a hurricane is called the eye. It has very low pressure. Clouds rush toward it. But they start to spin due to Earth's rotation. As a result, the eye stays calm. It has no clouds and no wind. When people talk about the calm in the eye of the storm, that's what they mean.

Hurricanes also strike Australia. There they are called willy-willies. When they hit Asia, they are called typhoons. Cyclones form in the Indian Ocean. It does not matter what they are called. They all act the same. These storms pull water from the sea. Then they dump it on land. Strangely, the water they drop isn't salty like sea water. That is because when the ocean water evaporates, it leaves its salt behind.

When it suddenly stops raining, becomes very calm, and the sun shines, you may think it's all over. But this is just the eye of the storm. The other side of the storm is about to hit! Hurricanes slowly die as they move inland. They leave a lot of damage in their wake.

High winds are not the only problem. The storm pushes water in front of it. As this water reaches land, it piles up in a wave called a storm surge. This causes huge flooding. The water level rises. Small buildings near the shore are suddenly underwater. Then, the water goes back to the ocean. It pulls things like cars and homes out to sea.

Scientists need to name hurricanes to keep track of them. For the Atlantic Ocean there are six lists of 21 names each. They switch to a new list each year. Boy and girl names go from A to W, skipping Q and U. If all 21 names are used, the Greek alphabet names the rest. This happened for the first time in 2005. There were a record number of hurricanes.

So, what sorts of names are picked? Here is the list for the year 2005: Arlene, Bret, Cindy, Dennis, Emily, Franklin, Gert, Harvey, Irene, Jose, Katrina, Lee, Maria, Nate, Ophelia, Philippe, Rita, Stan, Tammy, Vince, Wilma, Alpha, Beta, Gamma, Delta, Epsilon, and Zeta. Was your name a hurricane that year?

Comprehension Question

Compare hurricanes and tornadoes.

Tornadoes and Hurricanes

A tornado acts like a gigantic vacuum cleaner. It roars along at a speed of about 50 kilometers (30 miles) per hour. It smashes everything in its path. It can even lift cars into the air, dropping them far away.

Tornadoes can happen anywhere. The central states in the United States have more tornadoes than anywhere else on Earth. Each year about 1,000 tornadoes occur. Luckily, most are weak. Others are strong. They can do a great deal of damage.

Tornadoes can begin as thunderstorms. Thunderstorms can form supercells. A supercell is a large thunderstorm that rotates. A supercell can be very dangerous. Supercells can cause hail, damaging winds, and tornadoes.

A supercell forms when strong winds blow in opposite directions. This starts when the air within a thunderhead begins to spin. If it touches the ground, it becomes a tornado.

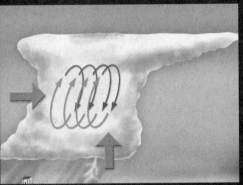

A thunderstorm can develop a supercell inside it. Winds go up, over, down, and back in a circle.

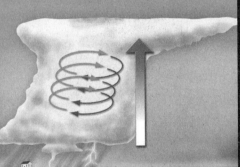

The supercell can get tilted up and down by a strong updraft.

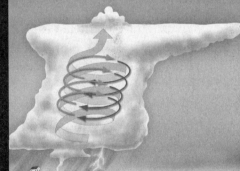

If the vertical supercell keeps going, it starts sucking wind and everything else up the cell. A tornado is born.

Hurricanes

As bad as tornadoes are, they aren't the worst storms. Each year hurricanes cause more damage than all other storms combined. Why? Hurricanes are huge, and they last a long time. Hurricanes start as tropical storms. They form in late summer or fall over warm water. The center of a hurricane is called the eye. It has very low pressure. Clouds rush toward it. But they start to spin due to Earth's rotation. As a result, the eye stays calm. It has no clouds and no wind. When people talk about the calm in the eye of the storm, that's what they mean.

Hurricanes also strike Australia, where they are called willy-willies. When they hit Asia, they are called typhoons. Cyclones form in the Indian Ocean. No matter what they are called, they act the same. These storms gather water from the sea. Then, they dump it on land. Strangely, the water they drop isn't salty like sea water. That is because when the ocean water evaporates, it leaves its salt behind.

When it suddenly stops raining, becomes very calm, and the sun shines, you may think it's all over. But this is just the eye of the storm. The other side of the storm is about to hit! Hurricanes slowly die as they move inland. They leave tremendous damage in their wake.

High winds are not the only problem. The storm pushes water in front of it. As this water reaches land, it piles up in a wave called a storm surge. This causes huge flooding. The water level rises. Small buildings near the shore are suddenly underwater. Then, the water recedes. It pulls things like cars and homes out to sea.

Scientists name hurricanes in the Atlantic Ocean by using six lists of 21 names each. They switch to a new list each year. Boy and girl names alternate from A to W, skipping Q and U. If all 21 names are used, the Greek alphabet names the rest. This happened for the first time in 2005. There were a record number of hurricanes.

So, what sort of names are picked? Here is the list for the year 2005: Arlene, Bret, Cindy, Dennis, Emily, Franklin, Gert, Harvey, Irene, Jose, Katrina, Lee, Maria, Nate, Ophelia, Philippe, Rita, Stan, Tammy, Vince, Wilma, Alpha, Beta, Gamma, Delta, Epsilon, and Zeta. Was your name a hurricane that year?

Comprehension Question

Compare and contrast hurricanes and tornadoes.

Tornadoes and Hurricanes

A tornado acts like a gigantic vacuum cleaner roaring along at a speed of about 50 kilometers (30 miles) per hour. It smashes everything in its path. It can even lift cars into the air, dropping them far away.

Tornadoes can happen anywhere, although the central states in the United States have more tornadoes than anywhere else on Earth. Each year about 1,000 tornadoes occur. Luckily, most are weak. It is the few strong hurricanes that do the most damage.

Tornadoes can begin as thunderstorms that form supercells. A supercell is a large, rotating system of winds. A supercell can be very dangerous. It can cause hail, damaging winds, and tornadoes.

A supercell forms when strong winds blow in opposite directions, causing the air within the thunderhead to spin. If it touches the ground, it becomes a tornado.

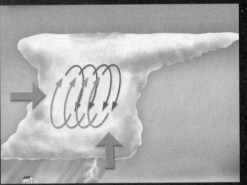

A thunderstorm can develop a supercell inside it. Winds go up, over, down, and back in a circle.

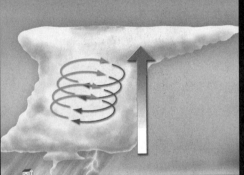

The supercell can get tilted up and down by a strong updraft.

If the vertical supercell keeps going, it starts sucking wind and everything else up the cell. A tornado is born.

Hurricanes

As destructive as tornadoes are, they aren't the worst storms. Each year hurricanes cause more damage than all other storms combined. Hurricanes are huge and persist for months at a time. Hurricanes start as tropical storms, forming in late summer or fall over warm water. The low-pressure center of a hurricane is called the eye. The clouds that rush toward it start to spin due to Earth's rotation, and as a result, the eye stays calm: it has no clouds and no wind. When people talk about the calm in the eye of the storm, that's what they mean.

Hurricanes also strike Australia, where they are called willy-willies; when they hit Asia, they are called typhoons; cyclones form in the Indian Ocean. No matter what they are called, they act the same. These storms gather water from the sea and dump it on land. Strangely, the water that hurricanes drop isn't salty like seawater because when the ocean water evaporates, it leaves its salt behind.

When it suddenly stops raining, becomes very calm, and the sun shines, you may think it's all over, but this is just the eye of the storm. The other side of the storm is about to hit! Hurricanes slowly die as they move inland, and leave tremendous damage in their wake.

High winds are not the only problem: the storm pushes a massive amount of water in front of it. As this water reaches land, it piles up in a wave called a storm surge. This causes huge flooding, drastically raising the water level. Small buildings near the shore are suddenly underwater. Then, the water recedes, pulling cars and homes out to sea.

Scientists name hurricanes in the Atlantic Ocean by using six lists of 21 names each. They switch to a new list each year, and alternate boy and girl names from A to W, skipping Q and U. If all 21 names are used, the Greek alphabet is used for the remainder. This happened for the first time in 2005, when there was a record number of hurricanes.

So, what sort of names are picked? Here is the list for the year 2005: Arlene, Bret, Cindy, Dennis, Emily, Franklin, Gert, Harvey, Irene, Jose, Katrina, Lee, Maria, Nate, Ophelia, Philippe, Rita, Stan, Tammy, Vince, Wilma, Alpha, Beta, Gamma, Delta, Epsilon, and Zeta. Was your name a hurricane that year?

Comprehension Question

Describe the important differences between tornadoes and hurricanes.

Structure of the Earth

Have you ever wondered how Earth is put together? Most people live their lives every day. They don't think about what is under their feet. Some people wonder about Earth. They think about Earth. They explore Earth. They are geologists. They study Earth. They study its structure. They have learned many things.

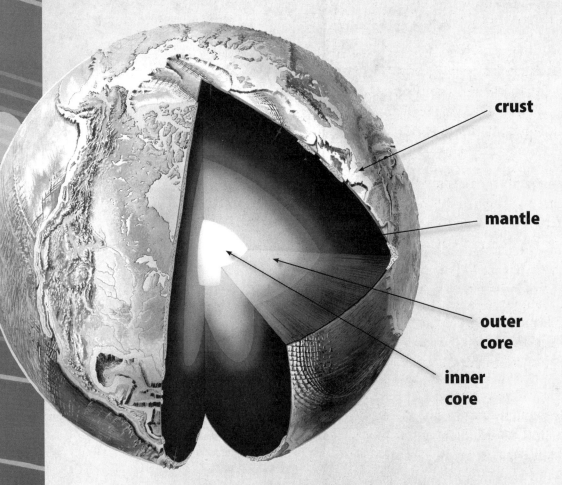

crust
mantle
outer core
inner core

For Eggsample...

Think of Earth as a hard-boiled egg. An egg has a shell. Earth has a crust. An egg has liquid under its shell. Earth has hot magma under its crust. The Earth is much bigger than an egg. It is 6,400 kilometers (4,000 miles) from its crust down to its center!

We live on Earth's crust. The crust is the part of Earth that has cooled and hardened. There are seven continents on Earth. They are all a part of the crust. The ocean floor is also a part of the crust. Mountains rise up from the crust.

An eggshell can get a crack. Our crust is cracked. Look at the edges of the continents. You might notice that they look like cracks on an eggshell.

Earth also has other layers under the crust. The first layer is the mantle. Then comes the core. There is an outer core. Then there is an inner core. These layers are made of magma, or molten rock. The heat there can go up to hundreds and thousands of degrees.

Recycling Crust

Magma comes to the surface. It comes through cracks in Earth's crust. This makes new crust. Does that mean there is more crust now than in the past? That doesn't make sense. Geologists have worked hard to find out how it works.

Earth oozes magma in one place. It makes new crust there. For this to happen it must destroy crust some other place. Geologists looked for that place. They found how it works. New crust is made in the Atlantic Ocean. It is destroyed in the Pacific. The Atlantic Ocean floor expands with new crust. The Pacific Ocean floor shrinks.

Geologists found out how. The Pacific Ocean floor dives down. It goes into deep trenches. These are called subduction zones. The crust in the ocean floors shows how Earth recycles. Rocks are created. Later they are melted again. Proof of this comes from maps. The maps show where there have been earthquakes. They show where there are volcanoes. The maps show undersea ridges and subduction zones.

Comprehension Question

Name the layers of Earth.

Structure of the Earth

Have you ever wondered how Earth is put together? Most people live their lives every day without thinking about the planet under their feet. Some people do. They wonder about Earth. They think about it. They explore it. These people are geologists. They study Earth and its structure. They have discovered many things.

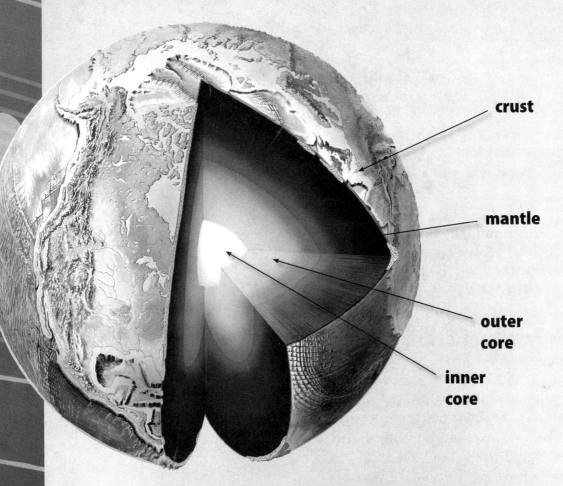

For Eggsample...

Think of Earth as a hard-boiled egg. An egg has a shell. Earth has a crust. An egg has liquid under its shell. Earth has hot magma under its crust. The Earth is much bigger than an egg. It is 6,400 kilometers (4,000 miles) from its crust down to its center!

We live on Earth's crust. The crust is the part of Earth that has cooled and become hard. All of the continents of Earth are a part of the crust. The ocean floor is also a part of the crust. Mountains rise up from the crust.

Just like an eggshell with a crack, our crust is cracked. Look at the edges of the continents. You might notice that they look like cracks on an eggshell.

Earth also has other layers beneath the crust. The first layer is the mantle. Then come the outer and inner cores. These layers are made of magma, or molten rock. Their temperatures range from hundreds to thousands of degrees Fahrenheit.

Recycling Crust

Molten magma rises to the surface through cracks in Earth's crust. This makes new crust. Does that mean there is more crust on the surface of Earth now than in the past? That doesn't make sense. Geologists had a theory to explain what happens.

Earth oozes magma in one place. It makes new crust there. It must destroy crust somewhere else. Sure enough, studies found how it works. New crust is made in the Atlantic Ocean, and is destroyed in the Pacific Ocean. The Atlantic Ocean floor expands with new crust. The Pacific Ocean floor shrinks.

Geologists found out how. The Pacific Ocean floor dives down. It goes into deep trenches under continents. These trenches are called subduction zones. The crust in the ocean floors is an example of how Earth recycles. Rocks are created and later melted again. Proof of this comes from mapping earthquakes and volcanoes. Most of them are found near undersea ridges and subduction zones.

Comprehension Question

Describe the layers of Earth.

Structure of the Earth

Have you ever wondered how Earth is put together? Most people live their lives every day without thinking about the planet under their feet. Some people do: they wonder about Earth; they investigate it, they theorize about it, and they explore it. These people are geologists, and they study the Earth and how it is structured. They have discovered a number of interesting things.

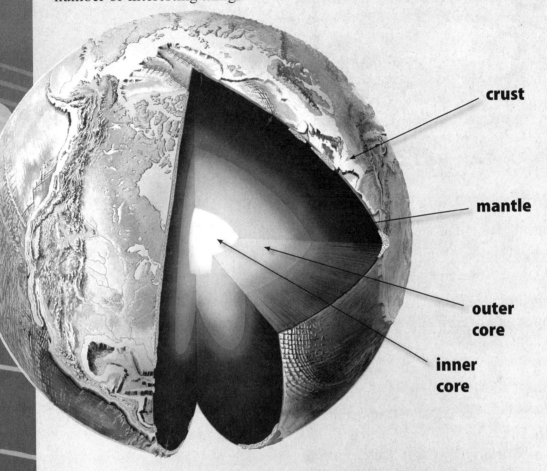

For Eggsample...

Imagine Earth as a hard-boiled egg. An egg has a shell. Earth has a crust. An egg has liquid under its shell. Earth has hot magma under its crust. If Earth were an egg, it would be a 6,400-kilometer (4,000-mile) trip from its shell (the crust) down to its center!

We live on Earth's crust. The crust is the part of Earth that has cooled and hardened. All of the continents of Earth are a part of the crust. The ocean floor is also a part of the crust. Mountains rise up from the crust.

Just like an eggshell with a crack, our crust is cracked. If you look at the edges of the continents, you might notice that they look like cracks on an eggshell.

Earth also has other layers beneath the crust. They are the mantle and the outer and inner cores. These layers are made of magma, or molten rock, and their temperatures range from hundreds to thousands of degrees Fahrenheit.

Recycling Crust

Molten magma rises to the surface through cracks in Earth's crust. This makes new crust. Does that mean there is more crust on the surface of Earth now than in the past? That doesn't make sense. Geologists had a theory to explain the phenomenon.

If Earth oozed molten magma in one place, then it must reabsorb crust somewhere else. Sure enough, studies reveal that the Atlantic Ocean floor is expanding and the Pacific Ocean floor is shrinking. New crust is made in the Atlantic Ocean, and is destroyed in the Pacific.

Geologists found that the Pacific Ocean floor dives down into deep trenches under continents. These trenches are called subduction zones. The expanding and shrinking ocean floors are an example of how Earth is really a recycler. Rocks are created and later recycled. Proof of recycling rocks comes from mapping earthquakes and volcanoes. Most of them are found near undersea ridges and subduction zones.

Comprehension Question

Describe how the layers of the Earth interact.

Structure of the Earth

Have you ever wondered how Earth is put together? Most people live their lives every day without thinking about the planet under their feet. Some people do: they wonder about Earth; they investigate it, they theorize about it, and they explore it. These people are geologists, and they study Earth and how it is structured. They have discovered a number of interesting things.

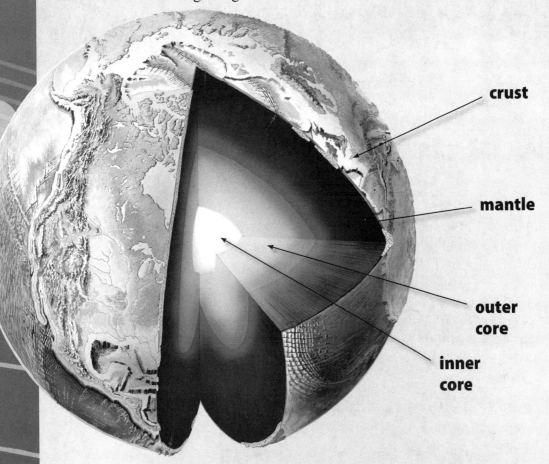

For Eggsample...

Imagine Earth as a hard-boiled egg: an egg has a shell and Earth has a crust; an egg has liquid under its shell and Earth has hot magma under its crust. If Earth were an egg, it would be a 6,400-kilometer (4,000-mile) trip from its shell (the crust) down to its center!

We live on Earth's crust, the cooled and hardened outer shell of the planet. All of the continents of Earth are a part of the crust, and so is the ocean floor.

Just like a cracked eggshell, Earth's crust is cracked into multiple pieces. If you look at the edges of the continents, you might notice that they look like cracks on an eggshell.

Earth also has other layers beneath the crust. They are the mantle and the outer and inner cores. These layers are made of magma, or molten rock, and their temperatures range from hundreds to thousands of degrees Fahrenheit.

Recycling Crust

Molten magma rises to the surface through cracks in Earth's crust; when it cools, it creates new crust. That implies that there is more crust on Earth's surface today than there was millions of years ago. However, that couldn't be right, so geologists had a theory to explain the phenomenon.

If Earth oozed molten magma in one place, then it must reabsorb crust somewhere else. Sure enough, studies reveal that the Atlantic Ocean floor is expanding and the Pacific Ocean floor is shrinking. New crust is made in the Atlantic Ocean, and is destroyed in the Pacific.

Geologists found that the Pacific Ocean floor dives down into deep trenches under continents. These trenches are called subduction zones. The expanding and shrinking ocean floors are an example of how Earth is really a recycler. Rocks are created and later recycled. Proof of recycling rocks comes from mapping earthquakes and volcanoes. Most of them are found near undersea ridges and subduction zones.

Comprehension Question

Describe how material is recycled through the different layers of the Earth.

Earthquakes and Volcanoes

Earthquakes

There's nothing quite like getting caught up in an earthquake. The ground shakes and rolls. The Earth makes a big shift. Have you ever felt an earthquake? An earthquake causes Earth's surface to move and shift. A large earthquake can change the land in seconds.

The outer shell of Earth's crust is broken up into many parts. Those parts of the crust move. Stress builds between them. The places between the parts are called fault lines. When too much stress builds along a fault line, the rock breaks or suddenly shifts. This is an earthquake.

Types of Faults

There are three main types of faults. A normal fault runs at an angle to the ground. The stresses push out. They push away from the fault line. This causes one big piece of rock to drop. It goes under the other piece. The Rio Grande Valley in New Mexico is a normal fault.

A reverse fault is also when the fault line is at an angle. The stress is not the same. Here, the stresses push in toward the fault line. One big piece of rock moves up. It goes over the other part. A reverse fault can be found at Glacier National Park. A reverse fault is also called a thrust fault.

A strike-slip fault is the last kind. There are pieces of rock on each side of the fault. The pieces move sideways. They slip past each other. There is hardly any movement up or down. The San Andreas Fault in California is a strike-slip fault.

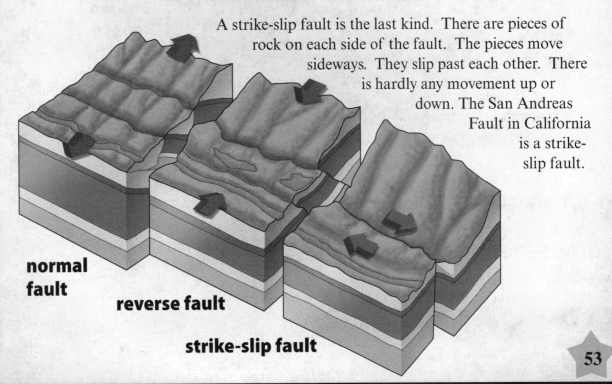

normal fault

reverse fault

strike-slip fault

Volcanoes Create Landforms

A volcano spits out melted rock, or magma. Magma comes from deep in the Earth. It pushes through where there is a weak spot in the crust. Most volcanoes happen near the edge of Earth's plates. The stress at the edges makes the crust weak. This lets magma burst out.

When the melted rock reaches the surface, it is called lava. Lava cools and gets hard. It becomes new landforms. Lava adds new rock to the land that is there. It can also form new islands.

Volcanoes have made many landforms on Earth. Mt. Fuji in Japan was made by volcanoes. So were the Hawaiian Islands. Volcanoes can also cause death and danger. Mt. St. Helens in the state of Washington, U.S.A., has done just that. It had a big blow up. The whole shape of the mountain was changed.

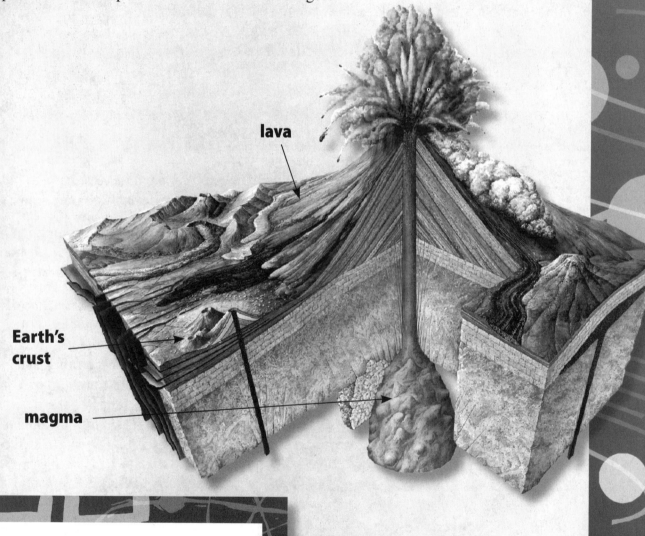

Comprehension Question

What causes earthquakes?

Earthquakes and Volcanoes

Earthquakes

If you live in some parts of the world, you know about earthquakes. There's nothing quite like getting caught up in one. The ground shakes and rolls. The Earth makes a big shift. Have you ever felt an earthquake? An earthquake causes Earth's surface to move and shift. A large earthquake can change the land in seconds.

The outer shell of Earth's crust is broken up into many pieces. As those pieces of the crust move, stress builds between them. The borders between pieces of crust are called fault lines. When too much stress builds along a fault line, the rock breaks or suddenly shifts. This is an earthquake.

Types of Faults

There are three main types of faults. A normal fault is when the fault line in Earth runs at an angle to the surface. The stress from an earthquake pushes out. The stress pushes away from the fault line. This causes one big piece of rock to drop below another piece. The Rio Grande Valley in New Mexico is a normal fault.

A reverse fault is also when the fault line is at an angle. The stress is not the same. Here, the stress from an earthquake pushes in toward the fault line. One big piece of rock moves up and over another piece. A reverse fault can be found at Glacier National Park. A reverse fault is also called a thrust fault.

A strike-slip fault is the last kind. The pieces of rock on each side of the fault move sideways. They do not move up and down as much. They slip past each other. There is little or no up-and-down movement. The San Andreas Fault in California is a strike-slip fault.

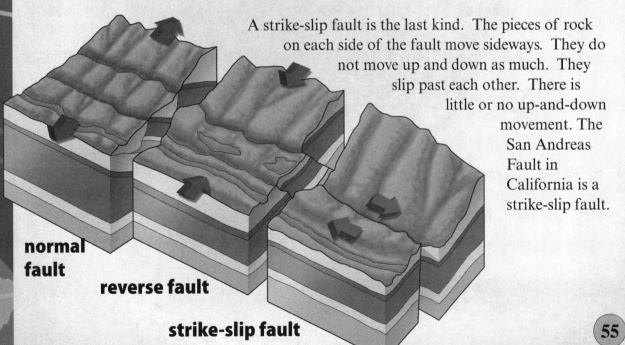

normal fault

reverse fault

strike-slip fault

Volcanoes Create Landforms

A volcano spits out melted rock, or magma. Magma comes from inside the Earth. It breaks through where there is a weak spot in the crust. Most volcanoes happen near the edge of Earth's plates. The push and pull at the edge makes the crust weak. This lets magma reach the surface.

When the melted rock reaches the surface, it is called lava. Lava cools and gets hard. It becomes new landforms. Lava adds new rock to the land that is there. It also forms new islands.

Volcanoes have made some of the most well known landforms on Earth. Mt. Fuji in Japan and the Hawaiian Islands were made by volcanoes. Volcanoes can also cause death and destruction. Mt. St. Helens in the state of Washington, U.S.A. has done just that. Due to the most recent big eruptions, the whole shape of the mountain has changed.

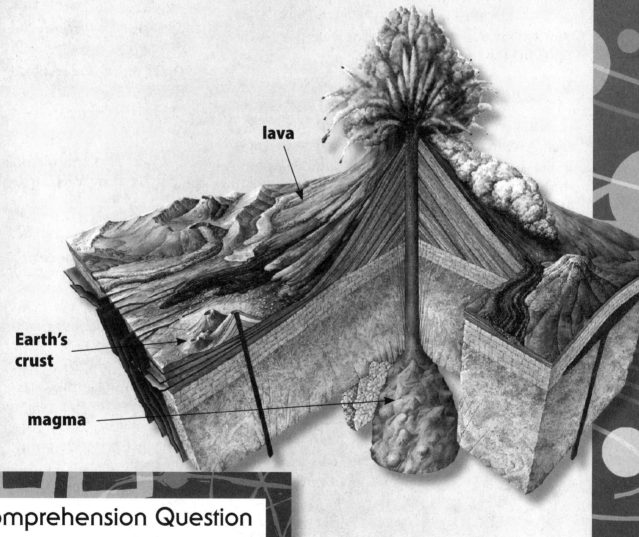

Comprehension Question

Why do earthquakes happen on fault lines?

Earthquakes and Volcanoes

Earthquakes

If you live in certain parts of the world, you are very familiar with earthquakes. There's nothing quite like getting caught up in all the shaking and rolling that goes on when the Earth makes a big shift. Have you ever felt an earthquake? An earthquake causes Earth's surface to move and shift. A large earthquake can change the land in seconds.

The outer shell of Earth's crust is broken up into many different pieces. As those pieces of the crust move, stress builds between them. The borders between pieces of crust are called fault lines. When too much stress builds along a fault line, the rock breaks or suddenly shifts. This is an earthquake.

Types of Faults

There are three main types of faults. A normal fault happens when the fault line in Earth runs at an angle to the surface. The stress from an earthquake pushes out, away from the fault line. This causes one section of rock to drop below another section. The Rio Grande Valley in New Mexico is an example of a normal fault.

A reverse fault also happens when the fault line is at an angle. But in this case, the stress from an earthquake pushes in toward the fault line. This causes one section of rock to move up and over another section. An example of a reverse fault can be found at Glacier National Park. A reverse fault is also called a thrust fault.

A strike-slip fault happens when the sections of rock on each side of the fault slip past each other sideways. There is little or no up-and-down movement. The San Andreas Fault in California is an example of a strike-slip fault.

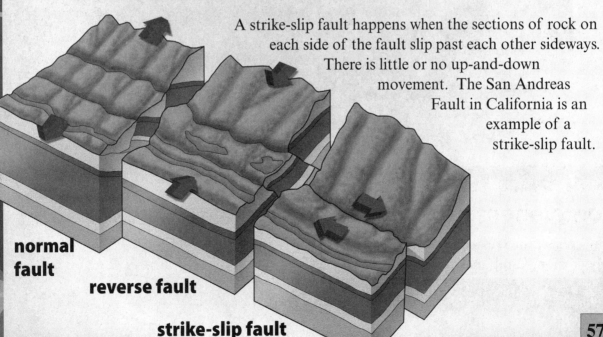

normal fault

reverse fault

strike-slip fault

Volcanoes Create Landforms

A volcano happens when melted rock, or magma, inside the Earth breaks through a weak spot in the crust. Most volcanoes happen near the edge of Earth's plates. The push and pull at these borders makes the crust weak. This lets magma reach the surface.

When the melted rock reaches the surface, it is called lava. Lava flowing from a volcano hardens to become new landforms. Lava adds new rock to existing land. It can also form new islands.

Volcanoes have made some of the most beautiful landforms on Earth. For example, Mt. Fuji in Japan and the Hawaiian Islands were made by volcanoes. Volcanoes can also cause death and destruction. Mount Etna in Italy has done just that. One of the most recent big eruptions occurred at Mt. St. Helens in the state of Washington, U.S.A. The shape of the mountain was changed completely.

Comprehension Question

Why do earthquakes and volcanoes happen along fault lines?

Earthquakes and Volcanoes

Earthquakes

If you live in certain parts of the world, you are very familiar with earthquakes. There's nothing quite like getting caught up in all the shaking and rolling that goes on when the Earth makes a big shift. In an earthquake, Earth's surface moves and shifts, and a large earthquake can change the land in seconds.

The outer shell of Earth's crust is broken up into many different pieces. As those pieces of the crust move, stress builds along the fault lines, or the borders between them. When too much stress builds along a fault line, the rock breaks or suddenly shifts, causing an earthquake.

Types of Faults

There are three main types of faults: normal, reverse, and strike slip. A normal fault happens when the fault line runs at an angle to the surface. The stresses from an earthquake push out, away from the fault line, causing one section of rock to drop below another section. The Rio Grande Valley in New Mexico is an example of a normal fault.

A reverse fault or thrust fault also happens when the fault line is at an angle, but in this case, the stress from an earthquake pushes in toward the fault line. This causes one section of rock to move up and over another section. An example of a reverse fault can be found at Glacier National Park.

A strike-slip fault happens when the sections of rock on each side of the fault slip past each other sideways. There is little or no up-and-down movement. The San Andreas Fault in California is an example of a strike-slip fault.

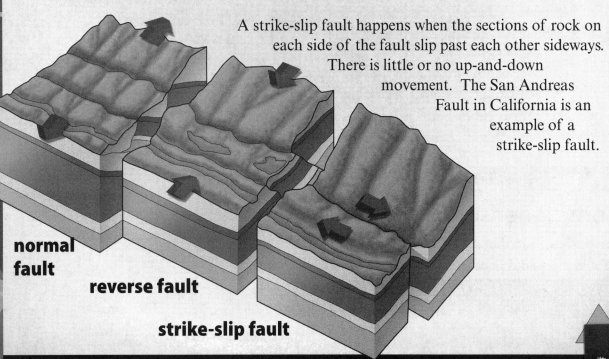

Volcanoes Create Landforms

A volcano happens when melted rock, or magma, inside the Earth breaks through a weak spot in the crust. Most volcanoes occur where pieces of Earth's crust meet, and where fault lines and earthquakes are common. The push and pull at these borders makes the crust weak, and magma can penetrate to the surface.

When the melted rock reaches the surface, it is called lava. Lava flowing from a volcano hardens and creates new landforms. Lava can add new rock to existing land, or it can even form completely new islands.

Volcanoes have made some of the most beautiful landforms on Earth. For example, Mt. Fuji in Japan and the Hawaiian Islands were made by volcanoes. Volcanoes can also cause death and destruction. Mount Etna in Italy has done just that. One of the most recent big eruptions occurred at Mt. St. Helens in the state of Washington, U.S.A. The shape of the mountain was changed completely.

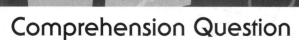

Comprehension Question

Describe the relationships between fault lines, earthquakes, and volcanoes.

Plate Tectonics

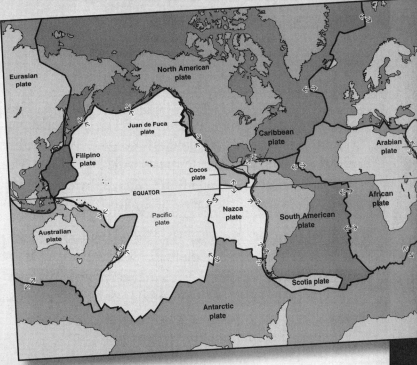

the tectonic plates of Earth

The surface of Earth is not one solid part. It is made of many parts. They fit side by side like a puzzle. Unlike a puzzle, those parts move. They push each other. They crash and smash. The parts are called tectonic plates. There are two types of plates on Earth. Oceanic plates are under the ocean water. Continental plates are under land.

The edges of plates meet at boundaries. There are three main types. They are divergent, transform, and convergent. Each kind works in its own way. Each kind can be found all over the world. They also make land features such as mountains and valleys.

Divergent Boundaries

Iceland is a small island. It is in the Atlantic Ocean. It is far to the north. It was made from a divergent boundary. It is on a ridge in the ocean. The ridge is where two plates meet. One plate goes west. The other plate goes east. They are very slow. They move at a rate of two to four centimeters per year (one inch per year).

Volcanoes are common there. The plates cause a gap in the land. The plates move. The gap gets bigger. Magma bursts up. It comes through Earth's crust. This makes a volcano. Over time, the magma piled up. It cooled down. It formed Iceland.

Transform Boundaries

Most transform boundaries are found in the ocean. The San Andreas Fault is on land. It is a transform boundary. You can find it in California. One side is the Pacific Plate. The other side is the North American Plate. They don't move apart. The plates slide past each other. This makes the plates grind. They slip and shake. This makes earthquakes.

Convergent Boundaries

Plates can form convergent boundaries. There are three ways. Each type has its own results.

Two ocean plates can smash. Then you get an ocean-ocean collision. That is what is at the Mariana Trench right now. The quick Pacific Plate smashes west. It hits the Filipino Plate. The first plate dives down. It goes under the second. It is melted. This makes earthquakes. It also makes volcanoes. The Mariana Islands are volcanoes. They started on the ocean floor. They have grown large. Now they peek out of the water. They are in the shape of an arc. That is the same shape as plates below them.

Then there are continent-continent collisions. Two plates smash head-on. They "fight it out." Then one plate loses. It subducts. It goes under the other. A lot of rock is scraped off the plate as it goes down. The rock piles up. It makes mountains. The Himalayas are the highest mountains in the world. They were made this way. It started 50 million years ago. The Indian and Eurasian plates crashed together. That formed the very tall mountain range.

The last kind to talk about is the ocean-continental collision. The ocean plate subducts. It dives under the other plate. There is one in South America right now. The ocean plate is diving under Peru and Chile. They have a lot of earthquakes. They even have volcanoes.

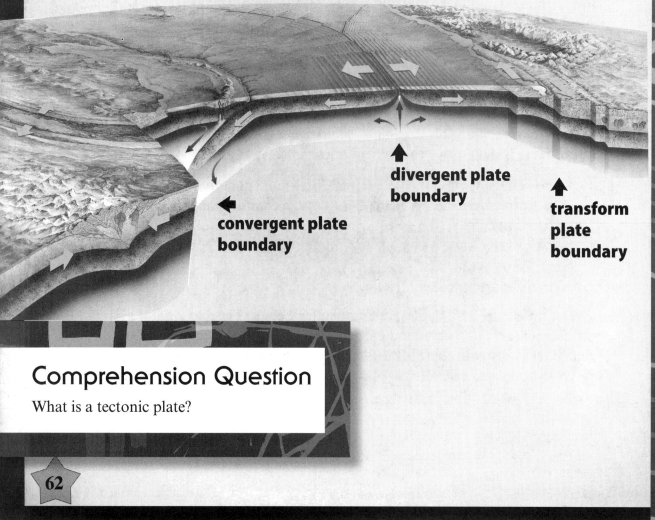

Comprehension Question

What is a tectonic plate?

Plate Tectonics

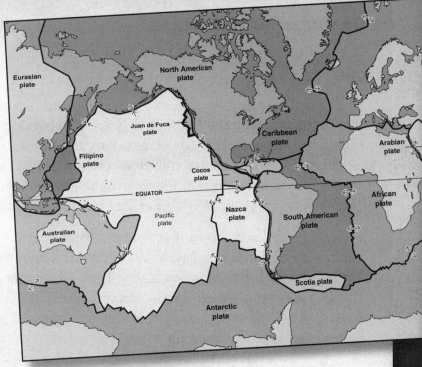

the tectonic plates of Earth

The surface of Earth is not one solid piece. It is made of many pieces. They fit side by side like a puzzle. Unlike a puzzle, those pieces move. They push each other. They crash and smash. The pieces are called tectonic plates. There are two types of plates on Earth. Oceanic plates are under the ocean water. Continental plates are under land.

The edges of plates meet at boundaries. There are three main types. They are divergent, transform, and convergent. Each kind acts in its own way. Each kind can be found all over the world. They also make land features such as mountains and valleys.

Divergent Boundaries

Iceland is a tiny island. It is in the Atlantic Ocean. You can find it between Norway and Greenland. It was made from a divergent boundary. Iceland is on a ridge in the middle of the ocean. Two plates are moving away from each other. They are very slow. They move at a rate of two to four centimeters per year (one inch per year).

Volcanoes are common on Iceland. The moving plates cause gaps. Magma bursts up and through Earth's crust. This action forms volcanoes. The cooled magma from the volcano formed Iceland.

Transform Boundaries

Most transform boundaries are found in the ocean. The San Andreas Fault is on land. The San Andreas Fault is a transform boundary. You can find it in California. One side is the Pacific Plate. The other side is the North American Plate. They don't move apart. They are sliding past each other. This causes major earthquakes.

Convergent Boundaries

Plates can form convergent boundaries in one of three ways. Each type has its own results.

When two ocean plates collide, you get an ocean-ocean collision. That is causing the Mariana Trench right now. The quick Pacific Plate is crashing into the Filipino Plate. The first plate dives down into Earth's mantle. It is melted. This causes earthquakes. It also causes volcanoes. The Mariana Islands are underwater volcanoes. They have grown large. They have grown high enough to peek out of the water. They are in the shape of an arc. That is the same shape as the ocean-ocean boundary.

Then there are continent-continent collisions. Two plates collide head-on. They "fight it out." Then one plate loses. It subducts under the other. A lot of rock is scraped off the plate as it goes down. The rock piles up. It makes mountains. The Himalayas are the highest mountains in the world. They were made this way. It started 50 million years ago. The Indian and Eurasian plates crashed together. That formed the very tall mountain range.

The last kind to talk about is the ocean-continental collision. The ocean plate subducts. It dives under the other plate. There is one in South America right now. It is diving under Peru and Chile. They have a lot of earthquakes. They even have volcanoes.

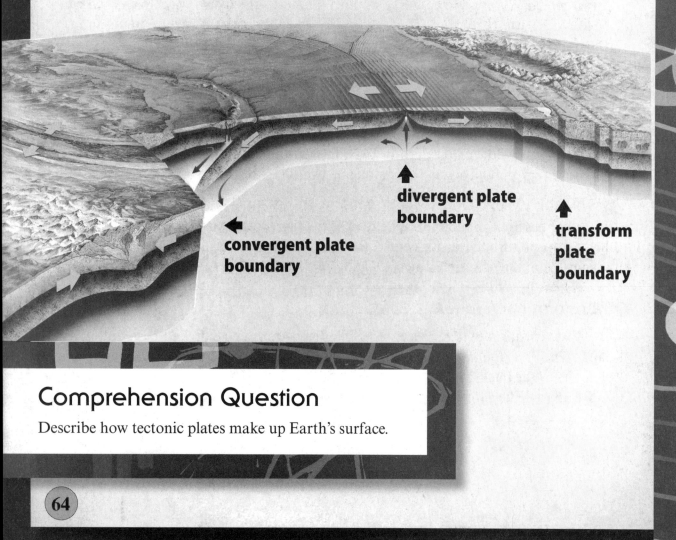

Comprehension Question

Describe how tectonic plates make up Earth's surface.

Plate Tectonics

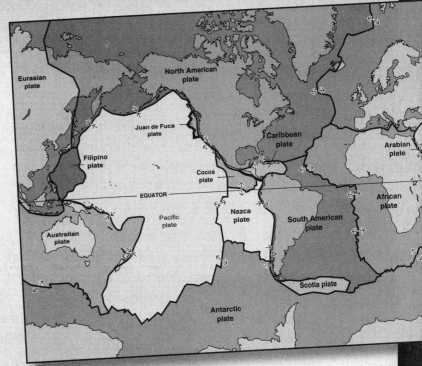

the tectonic plates of Earth

The surface of Earth is not one solid piece. Instead, it is made of many pieces that fit together like a puzzle. However, unlike a puzzle, those pieces move. They jostle against each other. Sometimes they crash and smash together. The pieces are called tectonic plates. There are two basic types of plates on Earth. Oceanic plates are under the ocean water. Continental plates make up the continents.

Scientists also know that plates have three main types of boundaries, or edges. They are divergent, transform, and convergent. Each kind behaves in a different way. The many boundaries can be found all over the world. They also make land features such as mountains and valleys.

Divergent Boundaries

Iceland is a tiny island. It is in the Atlantic Ocean. You can find it between Norway and Greenland. It was made from the divergent boundary of the midocean ridge. Two plates are moving away from each other very slowly. They move at a rate of two to four centimeters per year (one inch per year).

Volcanoes are common on Iceland. The movement of the plates causes gaps. Magma bursts up and through Earth's crust. This action forms volcanoes. The cooled magma from the eruptions formed Iceland.

Transform Boundaries

Most transform boundaries are found in the ocean. The San Andreas Fault is on land. The San Andreas Fault is a transform boundary. You can find it in California. It falls between the Pacific Plate and the North American Plate. These two plates aren't moving away from each other. They are sliding past each other. This sliding motion has caused major earthquakes.

Convergent Boundaries

Plates can form convergent boundaries in one of three ways. Each type has its own results.

An ocean-ocean collision involves two ocean plates. Right now, such a collision is causing the Mariana Trench. The fast-moving Pacific Plate is crashing into the Filipino Plate. The first plate dives into Earth's mantle. It is melted. This causes earthquakes. It also causes volcanoes. The Mariana Islands are underwater volcanoes. They have grown large enough to rise above the water line. They are in the shape of an arc. That is the same shape as the ocean-ocean boundary.

Then there is the continent-continent collision. Two plates collide head-on. They "fight it out." Then one plate loses and subducts under the other. A lot of rock is scraped off the plate as it subducts. The rock piles up to make mountains. The Himalayas are the highest mountains in the world. They are the result of a collision that started about 50 million years ago. The Indian and Eurasian continental plates crashed together. That formed the very tall mountain range.

The last kind to talk about is the ocean-continental collision. The oceanic plate subducts. It dives under the continental plate. There is one in South America right now. This is happening near Peru and Chile. They have a lot of earthquakes and volcanoes.

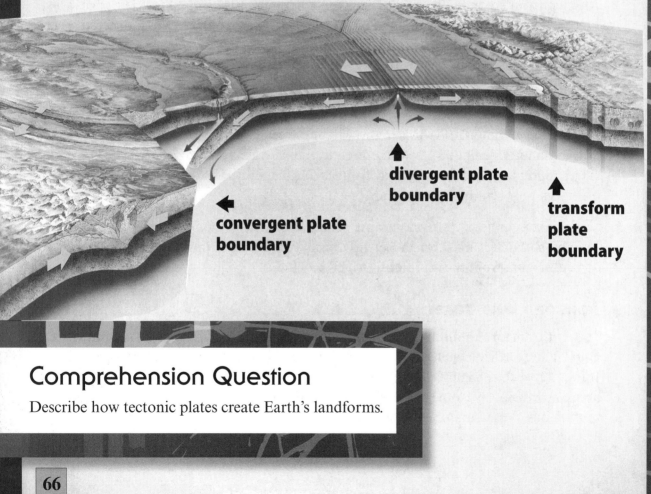

Comprehension Question

Describe how tectonic plates create Earth's landforms.

Plate Tectonics

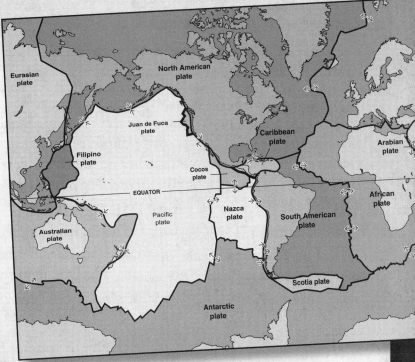

the tectonic plates of Earth

The surface of Earth is not one solid piece. Instead, it is made of many pieces that fit together like a puzzle. However, unlike a puzzle, those pieces move, jostling against each other and sometimes crashing and smashing together. The pieces are called tectonic plates. There are two basic types of plates on Earth. Oceanic plates are under the ocean water. Continental plates make up the continents.

Scientists also know that plates have three main types of boundaries, or edges. They are divergent, transform, and convergent. Each boundary behaves in a different way. The different boundaries can be found all over the world. The boundaries also make land features such as mountains and valleys.

Divergent Boundaries

Iceland is a tiny island in the Atlantic Ocean. It is between Norway and Greenland. It was made from the divergent boundary of the midocean ridge. Two plates are moving away from each other very slowly. They move at a rate of two to four centimeters per year (one inch per year).

Volcanoes are common on the island nation of Iceland. The movement of the plates causes magma to burst up and through Earth's crust. This action forms volcanoes. The cooled material from the volcanic eruptions formed Iceland.

Transform Boundaries

Most transform boundaries are found in the ocean, but the San Andreas Fault is on land. The San Andreas Fault in California is a transform boundary. It falls between the Pacific Plate and the North American Plate. These two plates are sliding past each other instead of pulling away from each other. This sliding motion has caused major earthquakes all along the state's coastline.

Convergent Boundaries

Plates can form convergent boundaries in one of three ways. Each type of convergent boundary has its own results.

An ocean-ocean collision happens between two ocean plates. Right now, such a collision is causing the Mariana Trench. The fast-moving Pacific Plate is crashing into the Filipino Plate. As the Pacific Plate dives into Earth's mantle, it is melted. This causes earthquakes and volcanoes. The Mariana Islands are underwater volcanoes that have grown large enough to rise above the water line. Such islands often form the shape of an arc. That is the same shape as the ocean-ocean collision boundary.

In a continent-continent collision, two plates collide head-on. They "fight it out" before one plate finally subducts under the other. A lot of material builds up as it is scraped off one plate as it subducts. The Himalayas are the highest mountains in the world. They are the result of a collision that started about 50 million years ago. The Indian and Eurasian continental plates crashed together to form the very tall mountain range.

An ocean-continental collision is happening in South America right now. An oceanic plate is being subducted under a continental plate. This is happening near Peru and Chile. That is why earthquakes and volcanoes are very common in this area of the world.

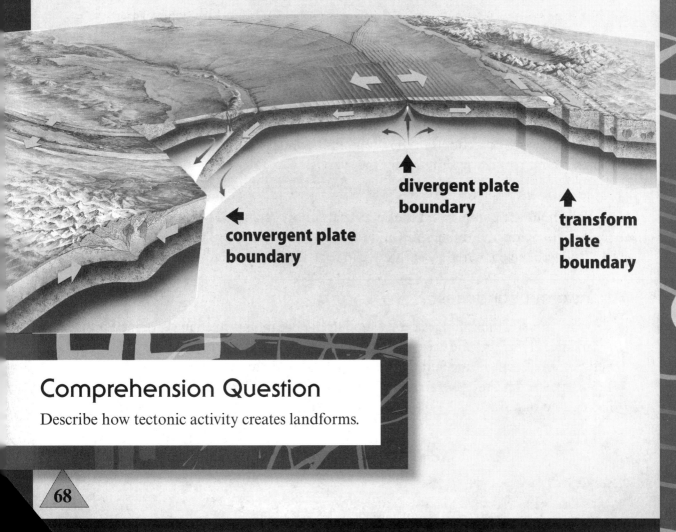

Comprehension Question

Describe how tectonic activity creates landforms.

Wegener Solves a Puzzle

Alfred Wegener was a scientist. He studied the weather. In his studies, Wegener saw something about Earth. Wegener looked at the coasts. He looked at the coast of South America. He looked at the coast of Africa. The coasts looked the same. He saw this on a map. It looked like a puzzle. He saw that much of Earth looked this way. It seemed like a giant jigsaw puzzle. It had pieces that fit next to each other.

Wegener's Clues

Wegener looked at fossils of animals. He did not think they fit the climate where they were found. The animals would freeze. He studied the weather. He knew all about climates. He looked at fossils from Antarctica. They were not built for cold. They must have come from a warm place close to the equator. Wegener said all of Earth's continents used to be in one mass. He called this first landmass Pangea. It means "all the earth." He was not the first to think this. He was the first to try to prove it. He used lots of sciences.

He looked at North America. He looked at Europe. He found the same fossils in both places. Some were on the east coast of North America. Some were on the west coast of Europe. He thought they were from Pangea. He thought they were split up when the landmass broke apart. He thought this was about 300 million years ago.

Next Wegener looked at South America. He looked at the east coast there. He looked at Africa. He looked at the west coast there. He found the same rocks in both places. He looked at the rocks from South America. They matched the rocks from Africa. This was good proof of Pangea.

Pangea

Scientists Argue

Geologists did not like his theory. Why? Wegener could not explain one part. How did all the continents move? He thought they moved through the ocean floor. He thought it was like ships. They smash through sheets of ice. He had been to the Arctic Ocean. He had seen ships do this in the icy waters. Most geologists did not think this would work. They knew that the ocean floor was thick and strong. They thought Wegener's theory was wrong. His ideas were ignored.

Wegener made good points. He showed geologists the puzzle pieces. He showed them the fossils. He showed them the rocks. Other geologists had their own explanation. They thought these things were caused by land bridges. They thought that land bridges once reached across the oceans. These land bridges were now sunk under the ocean.

Some scientists thought Wegener was right. One scientist used his work to study Africa and South America. He showed how they were the same in some ways. One more used his work. He was in the Swiss Alps. He used it in his own studies. Wegener did not live to see the world accept his theories. Today, this theory is called plate tectonics. It is part of how we think of the Earth.

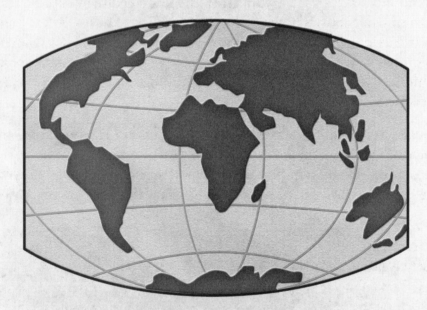

Present Day

Comprehension Question

What was Alfred Wegener's big idea?

Wegener Solves a Puzzle

Alfred Wegener was a scientist. He studied the weather. In his studies, Wegener noticed something about Earth. Wegener looked at coastlines. He looked at the coasts of South America and Africa. The coastlines fit together. He saw this while looking at a map. It looked like a puzzle. He saw that much of the Earth looked this way. It seemed like a giant jigsaw puzzle. It had pieces that fit next to each other.

Wegener's Clues

Wegener looked at fossils. He did not think those animals could have survived the climate there. He studied the weather and knew all about climates. He thought some fossils from Antarctica could not have survived the cold. They must have come from a place closer to the equator. Wegener said all of Earth's continents used to be connected. He called this single landmass Pangea. That means "all the earth." He was not the first person to suggest this. He was the first to try to prove it. He used many sciences.

He looked at North America and Europe. He found the same fossils in both places. Some were on the east coast of North America. Some were on the west coast of Europe. He thought the fossils were from Pangea. He thought they were separated when Pangea broke up. He thought this was about 300 million years ago.

Wegener also looked at South America and Africa. He found the same rocks in both places. He looked at the rocks on the east coast of South America. They matched the rocks of west Africa. This was good proof of Pangea.

Pangea

Scientists Argue

Geologists did not believe his theory. Why? Wegener could not explain how the continents moved. He figured they moved through the ocean floor. He thought it was like ships going through sheets of ice. He had been to the Arctic Ocean. He had seen ships do this in the icy waters. Most geologists thought that the continents could not move. They believed that the ocean floor was too thick and strong. They thought Wegener's theory was wrong. His ideas were rejected.

Wegener made good points. He showed geologists the puzzle pieces. He showed them the fossils. He showed them the rocks. They had their own explanation. They thought these things were caused by land bridges. They thought that land bridges once spanned the oceans. These land bridges were now sunk beneath the ocean.

Some scientists believed Wegener was right. One used Wegener's work. He studied Africa and South America. He explained how they were similar. One more used Wegener's work, too. He was in the Swiss Alps. He used it in his own studies. Wegener did not live to see the world agree with his theories. Today, this theory is called plate tectonics. It is widely accepted.

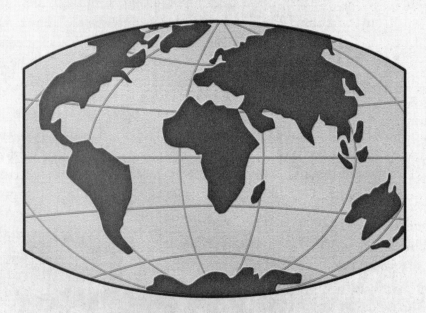

Present Day

Comprehension Question

What was Alfred Wegener trying to prove?

Wegener Solves a Puzzle

Alfred Wegener was a meteorologist. He studied the weather. In his studies, Wegener noticed something about Earth. Wegener saw that the coastlines of South America and Africa fit together. He saw this while looking at a map. It looked like a puzzle. He noticed that much of the Earth looked this way. It seemed like a giant jigsaw puzzle. It had pieces that fit together.

Wegener's Clues

Wegener thought that fossils found in some areas could not have survived the climate there. Remember, he was a meteorologist. He knew all about climates. He thought some fossils found in Antarctica could never have survived the cold. He figured they must have come from a place closer to the equator. Almost 100 years ago, Wegener said all of Earth's continents used to be connected. He called this single landmass Pangea. It means "all the earth." He was not the first person to suggest this. But he was the first to try to prove it. He used different sciences.

He looked at North America and Europe. He found the same plant and animal fossils in both places. Some were on the eastern coast of North America. Some were on the western coast of Europe. He figured that if the same fossils were in both places, then maybe those continents had been together at one time. He thought this happened about 300 million years ago.

Wegener also looked at South America and Africa. He found the same rocks in both places. He looked at the rocks on the eastern coast of South America. They matched the rocks of western Africa. This was good proof that Pangea had really existed.

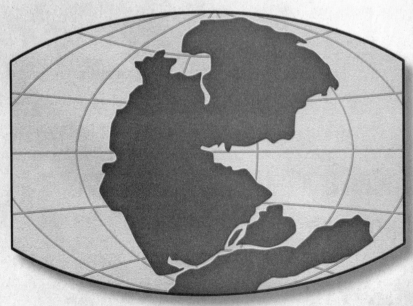

Pangea

Scientists Argue

Why did geologists not believe his theory? Wegener did not have a good explanation for how the continents moved. He figured that the continents moved through the ocean floor. He thought it was like ships going through sheets of ice. He had seen ships do this in the icy waters of the Arctic Ocean. Most geologists thought that the continents could not move. They believed that the ocean floor was too thick and strong. They thought Wegener's theory about how the continents move was wrong. His ideas were rejected.

Wegener made good points. He showed geologists the similar geological features, puzzle-piece shapes, and similar fossils. They had their own explanation. They thought these things were caused by land bridges. They thought that land bridges once spanned the oceans. These land bridges were now sunk beneath the ocean.

Some scientists supported Wegener. One used Wegener's ideas. He explained how Africa was like South America. Another used Wegener's ideas, too. He used them in his own observations in the Swiss Alps. Wegener didn't live to see the world accept his theories. Today, this theory is called plate tectonics. It is widely accepted.

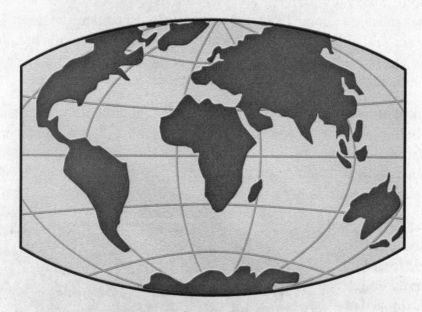

Present Day

Comprehension Question

Why was Alfred Wegener unable to prove his theory?

Wegener Solves a Puzzle

Alfred Wegener was a meteorologist. He studied the weather. In his studies, Wegener noticed something about Earth. Wegener saw that the coastlines of South America and Africa fit together. He saw this while looking at a map. It looked like a puzzle. He noticed that many of Earth's continents looked this way. They seemed like giant jigsaw puzzle pieces that could fit together.

Wegener's Clues

Wegener thought that fossils found in some areas could not have survived the climate there. Remember, he was a meteorologist. He knew all about climates. He thought some fossils found in Antarctica could never have survived the cold. He figured they must have come from a place closer to the equator. Almost 100 years ago, Wegener said all of Earth's continents used to be connected. He called this single landmass Pangea. It means "all the earth." He was not the first person to suggest this. But he was the first to try to prove it. He used different sciences.

He looked at North America and Europe. He found similar plant and animal fossils in both places. They were on the eastern coast of North America and the western coast of Europe. He figured that if the same fossils were in both places, then maybe those continents had been together at one time. He thought this happened about 300 million years ago.

Wegener also looked at South America and Africa. He found evidence that the rocks of the eastern coast of South America matched the rocks of western Africa. This was good proof that Pangea had really existed.

Pangea

Scientists Argue

Why did geologists not believe his theory? Wegener did not have a good explanation for how the continents moved. He figured that the continents moved through the ocean floor. He thought it was like ships going through sheets of ice. He had seen ships do this in the icy waters of the Arctic Ocean. Most geologists thought that the continents could not move. They believed that the ocean floor was too thick and strong. They thought Wegener's theory about how the continents move was wrong. His ideas were rejected.

Wegener made good points about the similar geological features, puzzle-piece shapes, and similar fossils. Many geologists had another explanation for these points. Their explanation was land bridges. It was thought that land bridges connected the continents at one time. These land bridges were now sunk beneath the ocean.

Some scientists supported Wegener. One geologist thought Wegener's ideas explained the similarities between Africa and South America. Another geologist believed that Wegener's ideas explained his own observations in the Swiss Alps. Wegener didn't live to see geologists accept his theories. Today, this scientific theory is called plate tectonics. It is widely accepted.

Present Day

Comprehension Question

Explain why Wegener's inability to prove his theory does not mean his theory is not true.

The Rock Cycle

There are many kinds of rocks. They have many shapes and sizes. They have many colors and textures. Rocks also have some things in common. They are natural. They are made of smaller parts. The parts are made of minerals. A mineral is a thing. Minerals can be man made. They are usually made by Earth. They can be made by living things on Earth. They are made of chemicals. Many of them form crystals.

A factory uses many steps to make things. They use heat, water, and force. They use machines to form their products. Earth is like a giant rock factory. No matter where you are, you could dig down deep into the earth. You would find rocks being made there.

Rocks can be sorted into three rock types. Some rocks are igneous. Some are sedimentary. The rest are metamorphic. The groups are sorted by how they are made.

Igneous

Igneous rocks form from magma. Magma is melted rock. It is liquid. It forms deep down inside the earth. There, the heat can reach thousands of degrees.

Some types of magma are thin and runny. They are like water. Other types are thick and gooey. They are like syrup. Magma gets pushed up to the surface of Earth. It squeezes through cracks and holes in solid rocks.

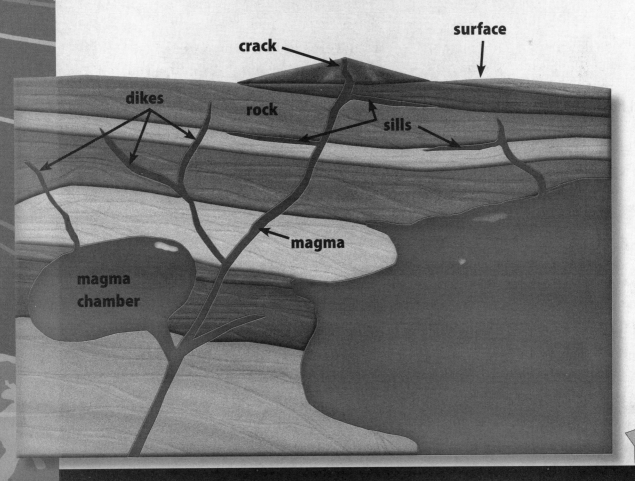

Sedimentary

On Earth's surface, rocks are changed by weather. The rocks are hit with rain and ice. They also get snow and wind. Rocks can be changed by chemicals. Plants and animals can also change rocks. Heat or cold can make them change, too.

These things cause rocks to break down. They fall apart into large and small pieces. Each piece is called a particle. The pieces can pile up in layers. That is called sediment. They are deposited as strata. Strata are layers of rock and soil in the earth.

1. Fast-moving water picks up rocks and soil.
2. The water slows down and deposits some rocks and soil.
3. Some sediment gets all the way to the ocean.

When water flows it can have lots of power. It can pick up these sediments. It moves them from place to place. Then the rock particles are dropped off. Other sediments can cover them. This is common in oceans and lakes. They can be buried under lots of layers of sediment. That much rock is very heavy. It puts a lot of pressure on the bottom. The particles get squeezed. They form new rocks. They are called sedimentary rocks.

Metamorphic

Deep down in the earth it is very hot. Things are smashed and pressed. These forces can change rocks. The rocks might melt. Then they turn back into magma. They might melt just halfway, too. Then they cool down. They are changed. They did not melt all the way. They are not igneous rocks. They are metamorphic rocks. That means that they have changed. Some examples of metamorphic rocks are schist, gneiss, and quartzite.

Comprehension Question

How are the three kinds of rocks made?

The Rock Cycle

There are many kinds of rocks. Rocks have many shapes and sizes. They have many colors and textures. Rocks also have some things in common. They are natural. They are made of smaller particles and minerals that are stuck together. A mineral is a thing found in nature. They are made by Earth. They can be made by living things on Earth. They are made of specific chemicals. Many minerals form crystals.

Factories use many processes to make things. They use heat, water, and force. They use machines to form their products. Earth is like a giant rock factory. No matter where you are, you could dig down deep into Earth. You would find rocks being made there.

Rocks can be divided into three rock types. They are igneous, sedimentary, and metamorphic. The groups describe how they are made.

Igneous

Igneous rocks form from a liquid called magma. Magma is melted rock. It forms deep beneath the earth's surface. There, the heat can reach thousands of degrees.

Some types of magma are thin and runny. They are like water. Other magmas are thick and gooey. They are like molasses syrup. Magma often gets pushed toward Earth's surface. It squeezes through cracks and holes in solid rocks.

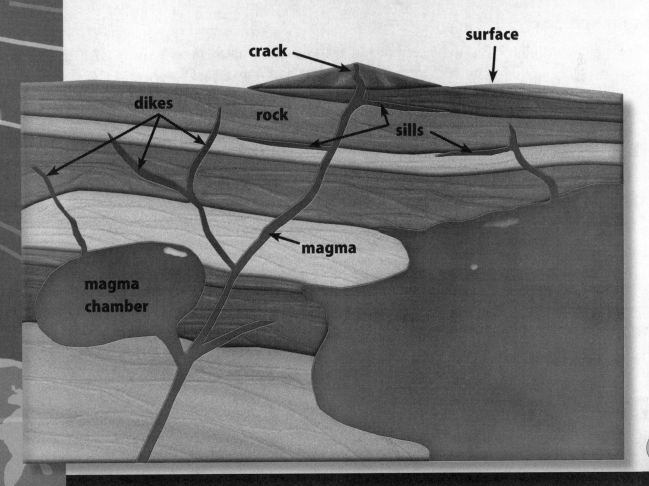

Sedimentary

At Earth's surface, rocks are changed by weather. The rocks are hit with rain, ice, snow, and wind. They can also be exposed to chemicals, plants, animals, and people. Extreme heat or cold can affect rocks, too.

These things cause rocks to break down. The rocks fall apart into large and small pieces. The pieces are called particles. Particles of broken rock are called sediment. They are deposited as strata. Strata are layers of rock and soil in the earth.

1. Fast-moving water picks up rocks and soil.

2. The water slows down and deposits some rocks and soil.

3. Some sediment gets all the way to the ocean.

Swiftly flowing water can pick up these sediments. It moves them to other places. Then the rock particles are dropped off. Other sediments can cover them. This is common in oceans and lakes. They can be buried under hundreds or even thousands of feet of sediment. That much rock is very heavy. It puts a lot of pressure on the lowest layers. The particles get squeezed together. They form new rocks. Rocks made in this way are called sedimentary rocks.

Metamorphic

Deep underground there is high pressure and heat that can transform rocks such as sandstone or granite. They might become liquid and turn into magma. They might melt halfway and then cool down. They become solid again. Since they did not melt all the way, they are not igneous rocks. They are metamorphic. That means that they have changed. Some examples of metamorphic rocks are schist, gneiss, and quartzite.

Comprehension Question

Describe the three kinds of rocks.

The Rock Cycle

There are many kinds of rocks. They come in a great variety of shapes, sizes, colors, and textures. Rocks also have some things in common. They are made naturally. They are made of smaller particles and minerals that are stuck together. Minerals are naturally occurring substances. They are made by Earth or organisms on Earth. They are made of specific chemicals. Many of them form crystals.

Factories use many processes to make things. They use heat, water, and force. They use machines to form their products. Earth is like a giant rock factory. No matter where you are, you could dig down deep into the earth. You would find rocks being made there.

Rocks can be divided into three rock types. They are igneous, sedimentary, and metamorphic. The groups describe the conditions that make them.

Igneous

Igneous rocks form from a liquid called magma. Magma is earth materials that have been melted. It usually forms deep beneath the earth's surface. There, the temperature is in the hundreds or thousands of degrees.

Some types of magma are thin and runny like water. Other magmas are thick and gooey like molasses. Magma often gets pushed toward the earth's surface. It squeezes through cracks and holes in solid rocks.

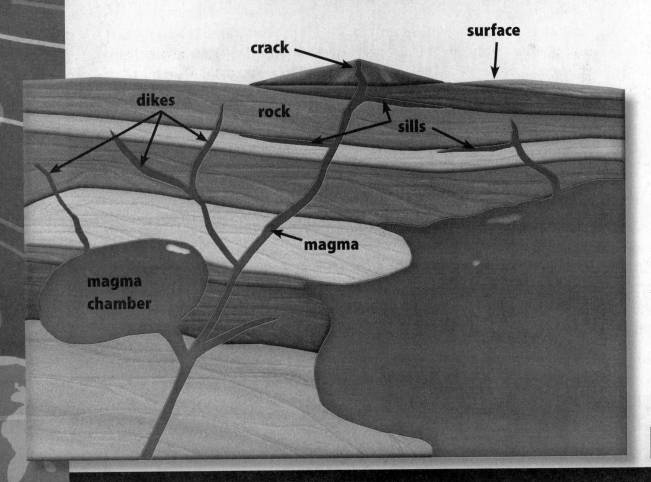

Sedimentary

On Earth's surface, rocks are changed by weather. They are hit with rain, ice, snow, and wind. They can also be exposed to chemicals, plants, animals, and people. Extreme heat or cold can affect rocks, too.

These conditions cause rocks to break down. They fall apart into large and small pieces, called particles. Particles of broken rock called sediment are deposited as strata. Strata are layers of rock and soil in the earth.

1. Fast-moving water picks up rocks and soil.
2. The water slows down and deposits some rocks and soil.
3. Some sediment gets all the way to the ocean.

Swiftly flowing water can pick up these sediments. It moves them to other places. After the rock particles are dropped off, other sediments can cover them. This is common in oceans and lakes. Over time, they can be buried under hundreds or even thousands of feet of sediment. That much sediment is very heavy. It puts great amounts of pressure on the lowest layers. The particles get squeezed together and form new rocks. Rocks made in this way are called sedimentary rocks.

Metamorphic

Deep underground there is high pressure and heat. They can transform rocks such as sandstone or granite. They might liquefy and turn into magma. They might melt halfway and then cool down. They become solid again. Since they did not melt all the way, they are not igneous rocks. They are metamorphic. That means that they have changed. Some examples of metamorphic rocks are schist, gneiss, and quartzite.

Comprehension Question

Compare and contrast the three kinds of rocks.

The Rock Cycle

There are many different kinds of rocks in an amazing variety of shapes, sizes, colors, and textures. Regardless of those differences, all rocks have some characteristics in common. Rocks are made naturally of groups of smaller particles and minerals that are stuck together. Minerals are naturally occurring substances that Earth or organisms on Earth produce. They form crystals and are made of specific chemicals.

Factories use different processes to make things. They use heat, water, and force from machines to form their products. Earth is like a giant rock factory. Wherever you are right now, if you could dig down far enough, you would find rocks being made deep inside the earth.

Rocks can be divided into three rock types: igneous, sedimentary, and metamorphic. These groups reflect the different conditions under which rocks are made.

Igneous

Igneous rocks form from Earth materials that have melted to a liquid called magma. Magma usually forms deep beneath the earth's surface where the temperature is in the hundreds or thousands of degrees.

Some types of magma are thin and runny like water. Other magmas are thick and gooey like molasses. Magma often gets pushed toward the earth's surface, where it squeezes through cracks and holes in solid rocks.

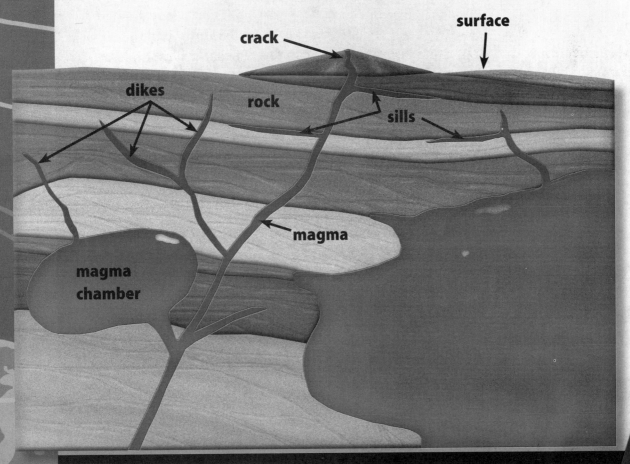

Sedimentary

At Earth's surface, rocks are affected by weather conditions. They are subjected to rain, ice, snow, and wind. They can also be exposed to chemicals, plants, animals, and people. Very hot or cold temperatures can affect rocks, too.

1. Fast-moving water picks up rocks and soil.

2. The water slows down and deposits some rocks and soil.

3. Some sediment gets all the way to the ocean.

These conditions cause rocks to break into large and small pieces, called particles. Particles of broken rock called sediment are deposited as strata. Strata are layers of rock and soil in the earth.

Sometimes, swiftly flowing water picks up these sediments and moves them to other places. When rock particles are dropped by water, other sediments can cover the particles. This often happens in oceans and lakes. After a long time, they can be buried under hundreds or even thousands of feet of sediment. That much sediment is very heavy, which puts great amounts of pressure on the lowest layers. There can be so much pressure that the particles get squeezed together and form new rocks. Rocks made in this way are called sedimentary rocks.

Metamorphic

Deep underground, high pressure or heat or both can force rocks such as sandstone or granite to transform. They might liquefy and turn into magma, or they might partially melt, cool down, and then become solid again. Because they are not melted completely, they don't become magma and are not igneous rocks. They are metamorphic, which means that they have changed. Some examples of metamorphic rocks are schist, gneiss, and quartzite.

Comprehension Question

Describe the products of the rock cycle.

Fun with Fossils

Have you ever found a rock that looks like a bone? Have you seen a rock with the print of a plant on it? These rocks are called fossils. They are made from living things. They have died and been buried. The buried bones and teeth can be turned into rock. It takes millions of years. Fossils can tell us about what kinds of things once lived on Earth. We know about plants and animals from long ago. We found their fossils. That's how we learned about dinosaurs. It's how we know about saber-toothed cats.

Ask a Fossil

Fossils are proof of past life. They are all that is left of living things from long ago. They can be leaf prints. They can be footprints. They can be shell prints and skeleton prints. Even the waste from living things can be turned to fossils! There are many kinds of fossils. They are made in many ways. They are made of many things. They can be made when a living thing is buried. They might be buried under ash from a volcano, mud, sand, or silt. They can be frozen in ice. They can turn into mummies from very dry air. Some fossils have been buried in tar for thousands of years.

Fossils are made from living things. Most fossils are not made from their soft parts. They are made when those parts rot. Then the hard parts are left. They turn into something like rock. The body is buried in dirt and soil. The minerals of the dirt and soil seep into the hard parts. They make the bone hard and strong. Fossils can be made of the whole living thing, too. It can be frozen whole. It can be turned into a mummy. Then the soft parts are included, too.

More fossils are made near water than on dry land. Near water, there is sand and silt. The living thing can get buried by these things. The silt piles up in layers. The layers turn into a kind of rock. It is called sedimentary. Fossils are often found in such rock.

What Fossils Tell Us

Think of fossils. Do you think of dinosaurs first? We know of many kinds of dinosaurs. We know about them through their fossils. Today there are no more dinosaurs. Those species have gone extinct. Most of what we know about them comes from fossils. This "fossil record" tells us many things. It tells us what kinds of animals were alive and when. It tells us where they lived. It tells us how they were shaped. Where we find fossils gives us clues. We can learn how those animals behaved and lived.

The fossil record tells us more. It does not just tell us about species that are gone. There are fossils of species that are still alive today. They also lived long ago. They lived when extinct species lived. That can tell us more about species that are alive now and that were alive in the past.

Comprehension Question

What is a fossil?

Fun with Fossils

Have you ever found a rock that looks like a bone? Have you found one that has the image of a plant on it? These rocks are called fossils. They are made when living things have died and are buried. Over millions of years, the buried bones and teeth can be turned into rock. Fossils can tell us about what kinds of things once lived on Earth. We know about plants and animals from long ago by finding their fossils. That's how we learned about dinosaurs and saber-toothed cats.

Ask a Fossil

Fossils are proof of past life. They are all that is left of living things from long ago. They can be leaf prints. They can be footprints. They can be shell prints and skeleton prints. Even the waste from living things can be turned to fossils! There are many kinds of fossils. They are made in many ways. They are made of many things. They can be made when a living thing is buried. They might be buried under ash from a volcano, mud, sand, or silt. They can be frozen in ice. They can turn into mummies from very dry air. Some fossils have been buried in tar for thousands of years.

Most fossils are not made from the soft parts of a living thing. They are made when these parts decay. Then the hard parts are turned into something like rock. The body is buried in dirt and soil. The minerals of the dirt and soil seep into the hard parts. They make the bone hard and strong. Fossils can be made when the whole living thing is frozen. They can be made when the whole thing is mummified. Then the soft parts are included, too.

More fossils are made near water than on dry land. Near water, the living thing will be buried by sand and silt. Over thousands of years, the silt piles up in layers. The layers turn into sedimentary rock. Fossils are often found in such rock.

What Fossils Tell Us

When you think of fossils, you might think of dinosaurs. Many kinds of dinosaurs have been found as fossils. Today there are no more dinosaurs. Those species have gone extinct. Most of what we know about dinosaurs comes from fossils. This "fossil record" tells us what kinds of animals were alive and when. It tells us where they lived. It tells us how they were shaped. Where we find fossils gives us clues about how those animals behaved and lived.

However, the fossil record tells us more. It doesn't just tell us about extinct species. There are fossils of species that are still alive today. They also lived millions of years ago. Some of today's species lived at the same time as species that are now extinct. Knowing that tells us more about species which are no longer with us.

Comprehension Question

How is a fossil made?

Fun with Fossils

Have you ever found a rock that looks like a bone? Maybe one that has the image of a plant on it? These rocks are called fossils. They are made when sediments quickly cover plants and animals that have died. Over millions of years, a buried animal's bones and teeth can be turned into rock. Fossils can tell us about what kinds of creatures once lived on Earth. We know about plants and animals from long ago by finding their fossils. That's how we learned about dinosaurs and saber-toothed cats.

Ask a Fossil

Fossils are proof of past life. They are the remains or imprints of living things from long ago. They can be leaf prints, footprints, shell prints, and skeleton prints. Even the waste from living things can become fossils! Different kinds of fossils are created through different processes and different materials. They can be made when a living thing dies and becomes buried. They might be buried under ash from a volcano, mud, sand, or silt. They can be frozen in ice. They can be mummified by very dry air. Some fossils have been buried in tar for thousands of years.

Most fossils are made when the soft parts of a living thing decay. The hard parts are turned into something like rock. The body is buried in sediment. The minerals of the sediments seep into the hard parts. They become preserved as fossils. In fossils that are made when the whole living thing is frozen or mummified, the soft parts are included, too.

Fossils are more likely to be made near a body of water than on dry land. Near water, it is likely to be buried quickly. Over thousands of years, the sediments settle into layers. They become sedimentary rock. Fossils are often found in such rock.

What Fossils Tell Us

When you think of fossils, you might think of dinosaurs. Tyrannosaurus, triceratops, and pterodactyls have been found as fossils. Today there are no more dinosaurs. Those species have become extinct. Most of what scientists know about dinosaurs comes from fossils. This "fossil record" tells us what kinds of animals were alive and when. It tells us where they lived. It tells us how they were shaped. Where we find fossils gives us clues about how those animals behaved and lived.

However, the fossil record doesn't just tell us about extinct species. There are fossils of species that are still alive today. They also lived millions of years ago. Some of today's species lived at the same time as species that are now extinct. Knowing that helps scientists understand more about species which are no longer with us.

Comprehension Question

How do scientists learn things from fossils?

Fun with Fossils

Have you ever found a rock that looks like a bone or that has the image of a plant on it? These rocks are called fossils, and they are made when sediments quickly cover organisms that have died. Over millions of years, a buried animal's bones and teeth can be turned into rock. Fossils can tell us a lot about what kinds of creatures once lived on Earth. We know about fascinating animals from long ago by finding their fossils. For example, that's how we learned about dinosaurs and saber-toothed cats.

Ask a Fossil

Fossils are evidence of past life: they are the remains or imprints of living things from long ago. They can be leaf prints, footprints, shell prints, and skeleton prints. The waste from living things can even become fossils! Different kinds of fossils are created through different processes and different materials. They can be made when a living thing dies and becomes buried by sediments, such as ash from a volcano, mud, sand, or silt. They can be frozen in ice or be mummified when the moisture is sucked up by very low-humidity air. Some fossils have been buried in tar for thousands of years.

Most fossils are made when the soft parts of a living thing decay; the hard parts are turned into something like rock. The minerals of the sediments in which the bodies are buried seep into the hard parts. They become preserved as fossils. In fossils that are made when the whole living thing is frozen or mummified, the soft parts are included, too.

Fossils are more likely to be made when a living thing dies near a body of water than on dry land. Near water, it is likely to be quickly buried. Over thousands of years, the sediments settle into layers that become sedimentary rock. Fossils are often found in sedimentary rock.

What Fossils Tell Us

When you think of fossils, you might think of dinosaurs. Tyrannosaurus, triceratops, and pterodactyls have been found as fossils. Today there are no more dinosaurs; those species have become extinct. Most of what scientists know about dinosaurs comes from fossils. This "fossil record" tells us what kinds of animals were alive and when, where they lived, and how they were shaped. Where we find fossils gives us clues about how those animals behaved and lived.

However, the fossil record is not limited to just extinct species. Scientists have found fossils of species that are still alive today, because those species were alive millions of years ago. Some present-day species coexisted with now extinct species, and knowing that helps scientists understand more about species which are no longer with us.

Comprehension Question

How does the creation of fossils leave clues about extinct species?

The Inner Planets

Our solar system is made of eight planets and the sun. A planet is a large body that orbits in space around a star. The sun is a star. It and the planets make up the solar system. The inner planets are close to the sun. They are all balls of rock. They all have thin layers of gas. Next to the other planets, they are very small. Our own Earth is one of these "small" planets.

The planets go around the sun. They also spin as they go. This is why we have day and night. When our part of Earth spins away from the sun, it is night. The time it takes for a planet to spin once is the planet's day. It takes Earth 24 hours to make one full spin. That means Earth's day is 24 hours long.

Mercury

Mercury is the closest planet to the sun. It is the smallest of the planets. It has lots of craters. These are shallow holes. The planet was struck by other space objects. They left craters. The planet can get very hot. Lead melts there! It gets as hot as 425 degrees Celsius (797 degrees Fahrenheit). But at night, it can get as cold as 150 degrees below zero Celsius (238 degrees below zero Fahrenheit). The big change is due to the long days there. One day on Mercury takes 59 Earth days. One night is that long, too. It can be very cold with no sunlight for 59 days. It can be very hot out in the sun for that long, too.

Venus

Venus is the second planet from the sun. It is the hottest planet. It's also the brightest in the night sky. Its thick clouds hold the sun's heat. It can get as hot as 462 degrees Celsius (864 degrees Fahrenheit)! Venus spins much slower than Earth. It takes 243 Earth days for it to spin once. It takes 225 Earth days for it to orbit the sun. That means that Venus's day is 18 Earth days longer than its year!

Life on Earth

Earth is not like any other planet we know of. It is the only planet to have life. It is the only one with oceans filled with water. We know that life needs water. The oceans make Earth look blue from space. Earth travels around the sun every 365 and $\frac{1}{4}$ days. That is one Earth year. As it orbits the sun, Earth spins at a speed of 1,660 kilometers (1,031 miles) per hour. In 24 hours, it spins once. That makes one Earth day.

Mars

Mars is a cold, windy planet. It has the largest mountains we know of. It has a big volcano. The volcano is more than three times taller than Earth's tallest mountain! Mars has a rusty color. That is why it is called the Red Planet. It is cold. It gets down to 60 degrees below zero Celsius (140 degrees below zero Fahrenheit). Water may have flowed on Mars long ago. There may have once been life there as well.

⬆ **This model shows a rover exploring Mars.**

Comprehension Question

How are the inner planets alike?

The Inner Planets

Our solar system consists of eight planets and the sun. A planet is a large body that orbits in space around a star. Our sun is a star. It and the planets make up the solar system. The inner planets are close to the sun. They are all balls of rock with thin layers of gas around them. Next to the other planets in the solar system, they are very small. Our own Earth is one of these "small" planets.

The planets orbit around the sun. They are also spinning. This is why we have day and night. When our part of Earth spins away from the sun, it is night. The time it takes for a planet to spin once is the planet's day. It takes Earth 24 hours to make one full spin. That means Earth's day is 24 hours long.

Mercury

Mercury is the closest planet to the sun. It is the smallest planet. Mercury has lots of craters. These are shallow holes. They were made when the planet was struck by other space objects. Mercury can get very hot. Lead melts there! It gets as hot as 425 degrees Celsius (797 degrees Fahrenheit). But at night, it can get as cold as 150 degrees below zero Celsius (238 degrees below zero Fahrenheit). The big range is due to the long days there. One day on Mercury takes 59 Earth days. One night is that long, too. It can be very cold with no sunlight for that long. It can be very hot out in the sun for that long, too!

Venus

Venus is the second planet from the sun. It is the hottest planet. It's also the brightest in the night sky. Its thick clouds hold the sun's heat. It can get as hot as 462 degrees Celsius (864 degrees Fahrenheit)! Venus spins much slower than Earth. It takes 243 Earth days for it to spin once. It takes 225 Earth days for it to orbit the sun. That means that Venus's day is 18 Earth days longer than its year!

Life on Earth

Earth is not like any other planet we know of. It is the only planet to have life. It is the only planet with oceans of water. Water is needed for life as we know it. The oceans make Earth look blue from space. Earth travels around the sun every 365 and $\frac{1}{4}$ days. That is one Earth year. As it orbits the sun, Earth spins at a speed of 1,660 kilometers (1,031 miles) per hour. In 24 hours, it spins once. That makes one Earth day.

Mars

Mars is a cold, windy planet. It has the largest mountains we know of. It has a big volcano that is more than three times taller than Earth's tallest mountain! Mars has a rusty color. That is why it is called the Red Planet. It is cold. It gets down to 60 degrees below zero Celsius (140 degrees below zero Fahrenheit). Water may have flowed on Mars long ago. There may even have once been life there.

⬆ **This model shows a rover exploring Mars.**

Comprehension Question

Compare the inner planets.

The Inner Planets

Our solar system has a number of planets orbiting around our sun. A planet is a large body that revolves in space around a star. Our sun is a star and its planets make up the solar system. The planets close to the sun are called the inner planets. They are all balls of rock with atmospheres of gas around them. Compared to the other planets in the solar system, they are very small. However, our own Earth is one of these "small" planets.

While the planets are revolving around the sun, they're also spinning, or rotating. This is why we have day and night. When our part of Earth rotates away from the sun, it's night. The time it takes for a planet to rotate once is the planet's day. It takes Earth 24 hours to make one rotation, so Earth's day is 24 hours long.

Mercury

Mercury is the closest planet to the sun. It is the smallest of the inner planets. Mercury is covered with craters. These are shallow holes. They were made when the planet was struck by other space objects. Mercury can get very hot. It can be hot enough to melt lead! It gets as hot as 425 degrees Celsius (797 degrees Fahrenheit). But at night, it can get as cold as 150 degrees below zero Celsius (238 degrees below zero Fahrenheit). The big range is due to Mercury's long days. One day on Mercury is about the same as 59 Earth days. One night is about that long, too. It can be very cold with no sunlight for 59 days, and it can be very hot so near to the sun for 59 more!

Venus

Venus, the second planet from the sun, is the hottest planet in our solar system. It's also the brightest planet in the night sky. Its thick clouds hold the sun's heat. It can get as hot as 462 degrees Celsius (864 degrees Fahrenheit)! Venus spins much more slowly than the Earth. It takes 243 Earth days for Venus to spin once. It takes 225 Earth days for Venus to orbit the sun. That means that Venus's day is 18 Earth days longer than its year!

Life on Earth

Earth is not like any other planet in our solar system. It is the only planet known to have life. No other planet has oceans of water. Water is needed for life as we know it. The oceans make it appear blue from space. Earth travels around the sun every 365 and $\frac{1}{4}$ days. That is one Earth year. As it orbits the sun, Earth is rotating at a speed of 1,660 kilometers (1,031 miles) per hour. In 24 hours, it makes one rotation: one Earth day.

Mars

Mars is a cold, windy planet. It has the largest mountains in the solar system. Its largest volcano is more than three times higher than Earth's tallest mountain! Mars is called the Red Planet because of its rusty color. It is cold. It gets down to 60 degrees below zero Celsius (140 degrees below zero Fahrenheit). Scientists think water may have flowed on Mars millions of years ago. They are still looking for signs that there may have been life there.

⬆ **This model shows a rover exploring Mars.**

Comprehension Question

Compare and contrast the inner planets.

The Inner Planets

Our solar system has a number of planets orbiting around the sun. A planet is an astronomical body that orbits in space around a star such as our sun. The inner planets, close to the sun, are all balls of rock with atmospheres of gas around them. Compared to the other planets in the solar system, they are significantly smaller. However, our own Earth is one of these "small" planets.

While the planets are orbiting around the sun, they're also rotating, creating day and night. Night is when our side of Earth rotates away from the sun. The time it takes for a planet to make one complete rotation is the planet's day. It takes Earth 24 hours to make one rotation, so Earth's day is 24 hours long.

Mercury

Mercury is the closest planet to the sun and is the smallest of the inner planets. Mercury is covered with craters, or shallow holes, that were made when the planet was struck by other space objects. Mercury can get very hot: hot enough to melt lead! The daytime temperature can rise to about 425 degrees Celsius (797 degrees Fahrenheit). At night, however, it can get as cold as 150 degrees below zero Celsius (238 degrees below zero Fahrenheit). The big range is due to Mercury's long days. One day on Mercury is about the same as 59 Earth days. One night is about that long, too. It can be very cold with no sunlight for 59 days, and it can be very hot so near to the sun for 59 more!

Venus

Venus, the second planet from the sun, is the hottest planet in our solar system. It's also the brightest planet in the night sky. Its thick clouds hold the sun's heat, bringing the temperature to 462 degrees Celsius (864 degrees Fahrenheit)! Venus spins much more slowly than Earth. It takes 243 Earth days for Venus to spin once. It takes 225 Earth days for Venus to orbit the sun. That means that Venus's day is 18 Earth days longer than its year!

Life on Earth

Earth is unlike any other planet in our solar system: it is the only planet known to have life. No other planet has oceans of water, and water is necessary for life as we know it. Earth's oceans make the planet appear blue from outer space. Earth travels around the sun every 365 and $\frac{1}{4}$ days: that is one Earth year. As it orbits the sun, Earth is rotating at a speed of 1,660 kilometers (1,031 miles) per hour. In 24 hours, it makes one rotation: one Earth day.

Mars

Mars is a cold, windy planet. It has the largest mountains in the solar system. Its largest volcano is more than three times higher than Earth's tallest mountain! Mars is called the Red Planet because of its rusty color. The average temperature is about 60 degrees below zero Celsius (140 degrees below zero Fahrenheit). Scientists think water may have flowed on Mars millions of years ago. They are still looking for signs that there may have been life there.

⬆ **This model shows a rover exploring Mars.**

Comprehension Question

What are the common attributes of the inner planets?

The Outer Planets

There are a many planets that orbit our sun. Some planets are like our own planet. Some are not like Earth. The planets of the outer solar system are strange. They are not like Earth at all!

Jupiter

Jupiter is the largest planet. You could put all the other planets together. It would still be bigger. It is one of four gas giants. They are planets made of gas. Jupiter has at least 63 moons. It has three rings, too. It has a huge storm in its clouds. It is called the Great Red Spot. The storm is at least 300 years old!

Saturn

Saturn is a very pretty planet. It has many rings. These rings are made of ice and rock. They are held in orbit around the planet. Their own speed pulls them out. Saturn's gravity pulls them in. They stay where they are. Saturn is the second largest of the outer planets. It is also made of gas. It has at least 59 moons. The most famous one is Titan. It is the only moon with air we know of.

Uranus

Uranus was found with a telescope. It was the first planet found this way. It looks like a blue green disk. It lies on its side. Some astronomers think a huge object may have crashed into it. It knocked Uranus on its side. *Voyager 2* is the only space probe to fly past the planet. It helped find most of its 27 known moons. Uranus has at least 11 rings.

Neptune

In 1846, scientists looked for one more planet. They knew Uranus moved strangely. They thought one more planet was the reason why. They made a chart of where they thought it might be. They pointed their telescopes at that point. There was Neptune!

Neptune is the last of the four giant planets. It is deep blue in color. It was named after the Roman god of the sea. It has six rings. It has at least 13 moons. The weather on Neptune is fierce. Winds whip up to more than 1,995 kilometers (1,250 miles) per hour!

Pluto

Scientists wanted to find a ninth planet. They tried the same trick. They charted where they thought it would be. When they looked, they found Pluto. It was found in 1930. It is far from the sun. It is 5.9 billion kilometers (3.7 billion miles) away! It takes Pluto 248 of our years to orbit the sun just once.

Pluto is tiny. It is just two-thirds the size of our moon. Is it a planet? Scientists did not know if it was big enough. They now call Pluto a dwarf planet.

Lowell Observatory

Comprehension Question

How is Pluto different from the outer planets?

The Outer Planets

There are many planets that orbit around our sun. Some planets are like our own planet. Some are not like Earth. The planets of the outer solar system are strange. They are not like Earth at all!

Jupiter

Jupiter is the largest planet in the solar system. In fact, it is more massive than the other seven planets combined! It is also one of four giant planets made of gases. Jupiter has at least 63 moons. It also has three faint rings. Its main feature is a huge storm called the Great Red Spot. This storm has lasted at least 300 years!

Saturn

Many people think Saturn is the prettiest planet. That's because of its thousands of rings. These rings are made up of ice and rock. They are held in an orbit around the planet. Their own speed pulls them out. Saturn's gravity pulls them in. That way they stay where they are. Saturn is the second largest planet in the solar system. It is made mostly of hydrogen and helium. It has at least 59 moons. The most famous one is Titan. It is the only moon in the solar system with air.

Uranus

Uranus was the first planet found with a telescope. It looks like a blue-green disk lying on its side. Some astronomers think a planet-sized object crashed into it. The crash may have knocked Uranus onto its side. *Voyager 2* was the only space probe to ever visit the planet. It helped find many of its moons. There are at least 27! Uranus has at least 11 rings around it.

Neptune

In 1846, scientists looked for one more planet. They knew Uranus moved strangely. They thought one more planet was the reason why. They made a chart of where they thought it might be. They pointed their telescopes at that point. There was Neptune!

Neptune is the last of the four giant planets. It was named after the Roman god of the sea. That's because of its deep blue color. It has six rings and at least 13 moons. The weather on Neptune is fierce. Winds whip up to more than 1,995 kilometers (1,250 miles) per hour!

Pluto

Scientists wanted to find a ninth planet. They tried the same trick. They charted where they thought it would be. When they looked, they found Pluto. It was found in 1930. It is far from the sun. It is 5.9 billion kilometers (3.7 billion miles) away! It takes Pluto 248 of our years to orbit the sun just once.

Pluto is tiny. It is just two-thirds the size of our moon. Is it a planet? Scientists did not know if it was big enough. They now call Pluto a dwarf planet.

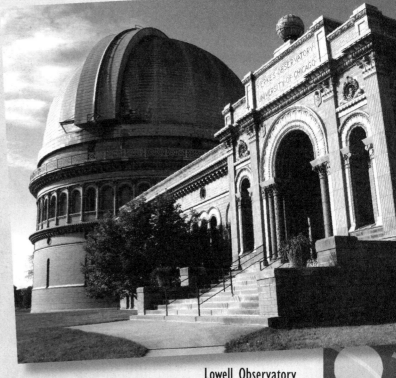

Lowell Observatory

Comprehension Question

What makes Pluto different from the outer planets?

The Outer Planets

There are a number of different planets orbiting around our sun. Some planets are like our own planet, but some are very different. The planets of the outer solar system are very different, indeed!

Jupiter

Jupiter is the largest planet in the solar system. In fact, it is more massive than the other seven planets combined! It is also one of four giant planets made of gases. Jupiter has at least 63 moons. It also has three faint rings. Its main feature is a huge storm called the Great Red Spot. This storm has lasted at least 300 years!

Saturn

Many people think Saturn is the most beautiful planet. That's because of its thousands of rings. These rings are made up of ice and rock. They are held in an orbit around the planet by their own speed and Saturn's gravity. Saturn is the second largest planet in the solar system. It is made mostly of hydrogen and helium. It has at least 59 moons. The most famous one is Titan. It is the only moon in the solar system with an atmosphere.

Uranus

Uranus was the first planet discovered with a telescope. It looks like a blue-green disk lying on its side. Some astronomers think a crash with a planet-sized object may have knocked Uranus onto its side. *Voyager 2* was the only space probe to ever visit the planet. It helped discover many of its 27 known moons. It's believed that Uranus has 11 rings rotating around it.

Neptune

In 1846, scientists went looking for a planet they thought might be moving Uranus with its gravity. They charted where they thought this new planet might be. When they pointed their telescopes at that point, there was Neptune!

Neptune is the last of the four giant planets. It was named after the Roman god of the sea. That's because of its deep blue color. It has six rings and at least 13 moons. The weather on Neptune is fierce. Winds whip up to more than 1,995 kilometers (1,250 miles) per hour!

Pluto

Scientists wanted to find a ninth planet. They tried the same trick as before. They charted where they thought it would be. When they looked, they found Pluto. It wasn't discovered until 1930. It is over 5.9 billion kilometers (3.7 billion miles) away from the sun. It takes Pluto 248 years to orbit the sun just once.

Pluto is tiny. It is just two-thirds the size of our moon. Scientists have argued whether it was big enough to be called a planet. They decided Pluto should be called a dwarf planet.

Lowell Observatory

Comprehension Question

What is the difference between a gas giant and a dwarf planet?

The Outer Planets

There are a number of different planets orbiting around our sun. Some planets are like our own planet, but some are very different. The planets of the outer solar system are very different, indeed!

Jupiter

Jupiter is the largest planet in the solar system; in fact, it is more massive than the other seven planets combined! It is also one of four giant planets made of gases. Jupiter has at least 63 moons, and it also has three faint rings. Its main feature is a huge storm called the Great Red Spot, which has lasted at least 300 years.

Saturn

Many people think Saturn is the most beautiful planet because of its thousands of rings. These rings are made up of ice and rock, and are held in an orbit around the planet by their own speed and Saturn's gravity. Saturn is the second largest planet in the solar system, but the least dense. That is because it is a giant mass of hydrogen and helium gas. Of its 59 moons, the most famous one is Titan. It is the only moon in the solar system with an atmosphere.

Uranus

Uranus was the first planet discovered with a telescope. It looks like a blue-green disk lying on its side. Some astronomers think a crash with a planet-sized object may have knocked Uranus onto its side. *Voyager 2* was the only space probe to ever visit the planet. It helped discover many of its 27 known moons. It's believed that Uranus has 11 rings rotating around it.

Neptune

In 1846, scientists went looking for a planet they thought might be moving Uranus with its gravity. They charted where they thought this new planet might be, and when they pointed their telescopes at that point, there was Neptune!

Neptune is the last of the four giant planets. It was named after the Roman god of the sea because of its deep blue color. It has six rings and at least 13 moons. The weather on Neptune is fierce: winds whip up to more than 1,995 kilometers (1,250 miles) per hour!

Pluto

In 1930, scientists wanted to find a ninth planet. They tried the same trick as before: they charted where they thought it would be, and when they looked there, they found Pluto. It is over 5.9 billion kilometers (3.7 billion miles) away from the sun. It takes Pluto 248 Earth years to orbit the sun just once.

Pluto is tiny, just two-thirds the size of our moon. Scientists argued whether it was big enough to be called a planet until they decided to call it a dwarf planet.

Lowell Observatory

Comprehension Question

Describe the differences between the gas giants and the dwarf planet Pluto.

Our Place in Space

Have you ever looked up at the night sky? Have you wondered what was out there? For all of time, people have looked up to the heavens. They have hoped to find clues about our place in the universe.

People used to think that Earth was the center of the solar system. They thought that the sun went around Earth. They thought the moon, stars, and other planets did, too. Then they learned that Earth is not the center. It is one of at least eight planets. All of the planets travel around the sun.

the Sun

Our Sun

The sun is a star. It is at the center of the solar system. It isn't the biggest star in the galaxy. It isn't the brightest. It is the star closest to Earth. It is the largest thing in our solar system. It has 99.8 percent of all of the mass in our solar system.

Most energy on Earth comes from the sun. It gives us light and heat which are used by plants to grow. It makes winds blow. It makes ocean currents flow. It makes the water cycle go.

With no sun, Earth would be very cold. It would be so cold that no living thing could survive.

The sun is made up of hot gases. Its center is about 15 million degrees Celsius (27 million degrees Fahrenheit). The sun also has a lot of gravity. Its gravity holds the planets close. It keeps Earth and other planets in place as they go around the sun.

The Four Seasons

Earth is tilted. That tilt causes the seasons. For part of the year, the northern half of Earth leans toward the sun. Then it gets lots of sunlight straight from the sun. It is summer there. At the same time, the southern half leans away. It gets less sunlight spread out over more land. It is winter there.

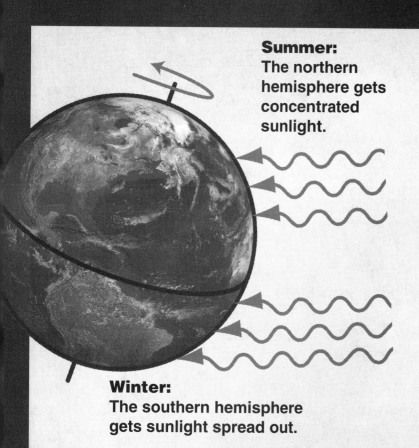

Summer: The northern hemisphere gets concentrated sunlight.

Winter: The southern hemisphere gets sunlight spread out.

The tilt does not change. The change comes from the Earth's orbit. Earth goes around the sun. As it goes, different parts get more light from the sun. That makes the seasons.

We also now know that our solar system is old. It is 5 billion years old. It is at the edge of our galaxy called the Milky Way. It is one of at least 100 billion galaxies in the heavens. Each one has tons of stars! The Milky Way has about 200 billion stars of its own.

The Milky Way is a spiral galaxy. It has many arms. Each arm is made of stars. You can see the next arm over. You can only see it on a very dark night. It looks like a bright stripe in the sky. The Milky Way is very wide. It takes light 100,000 years to get from one side to the other.

Comprehension Question

How is the Earth not the center of everything?

Our Place in Space

Have you ever looked up at the night sky? Have you wondered what was out there? For all of time, people have looked up to the heavens. They have hoped to find clues about our place in the universe.

People used to think that Earth was the center of the solar system. They thought that the sun went around the Earth. They thought the moon, stars, and other planets did, too. Then they learned that Earth is not the center. It is one of at least eight planets. All of the planets travel around the sun.

the Sun

Our Sun

The sun is a star. It is at the center of the solar system. It isn't the biggest or brightest star in our galaxy. The sun is the star closest to Earth. It is the largest body in our solar system. In fact, it has 99.8 percent of all of the mass in our solar system.

The sun is the major source of energy on Earth. It gives us light and heat. That energy is used for the growth of plants, winds, ocean currents, and the water cycle.

With no sun, Earth would be very cold. It would be so cold that no living thing could survive.

The sun is made up of hot hydrogen and helium gases. In its center it is about 15 million degrees Celsius (27 million degrees Fahrenheit). The sun also creates gravity. The force of gravity holds the sun and planets close. It keeps Earth and other planets in place as they orbit around the sun.

The Four Seasons

The way Earth is tilted causes the seasons. For part of the year, the northern half of Earth leans toward the sun. Then it gets direct sunlight. It is summer there. At the same time, the southern half leans away from the sun. It gets less sunlight. It is winter there.

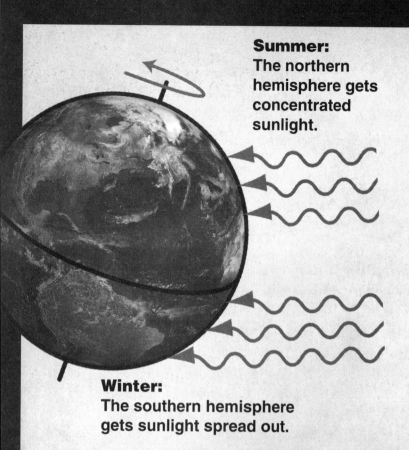

Summer: The northern hemisphere gets concentrated sunlight.

Winter: The southern hemisphere gets sunlight spread out.

The tilt doesn't change. It is the position of the planet around the sun that changes. As Earth goes around the sun, different parts get more sunlight. That makes the seasons.

We also now know that our solar system is old. It is almost 5 billion years old. It is at the edge of our galaxy called the Milky Way. It is one of at least 100 billion galaxies in the universe. Each one of them has tons of stars!

The Milky Way has about 200 billion stars on its own. The Milky Way is a spiral galaxy. It has many arms. The arms are made of stars. The next arm over can be seen on a very dark night. It looks like a bright band in the sky. The Milky Way is so wide that light would take 100,000 years to travel from one side to the other.

Comprehension Question

List three reasons why Earth is not the center of the universe.

Our Place in Space

Have you ever looked into the night sky and wondered what was out there? Throughout time, people have gazed to the heavens. They are hoping to find clues about our place in the universe.

Long ago, people thought that Earth was the center of the solar system. They thought that the sun, moon, stars, and other planets revolved around it. Scientists finally learned that Earth is one of at least eight planets in our solar system. They found that all of the planets travel around the sun.

the Sun

Our Sun

The sun is a star at the center of the solar system. It isn't the biggest or brightest star in our galaxy, but it is the star closest to Earth. It is the largest body in our solar system. In fact, it contains 99.8 percent of all of the mass in our solar system.

The sun is the major source of energy on Earth. It gives us light and heat. It's responsible for the growth of plants, winds, ocean currents, and the water cycle.

Without the sun, Earth would be very cold. It would be so cold that no living thing could survive.

Like other stars, the sun is made up of hot hydrogen and helium gases. The temperature in its center is about 15 million degrees Celsius (27 million degrees Fahrenheit). The sun also creates gravity. The force of gravity between the sun and planets keeps Earth and other planets of the solar system in place and orbiting around the sun.

Four Seasons

The way Earth is tilted causes the seasons. For part of the year, the northern half of Earth leans toward the sun and gets direct sunlight. So, it's summer there. At the same time, the southern half leans away from the sun and gets less sunlight. So, it's winter there.

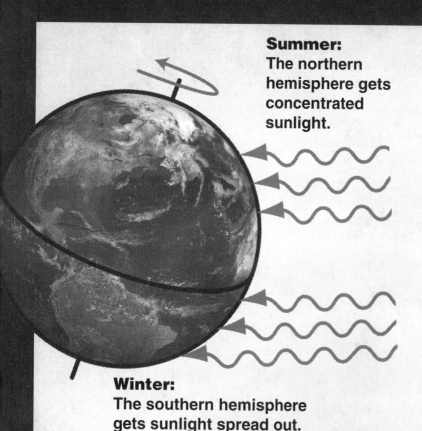

Summer: The northern hemisphere gets concentrated sunlight.

Winter: The southern hemisphere gets sunlight spread out.

The tilt doesn't change. It is the position of the planet around the sun that changes. As Earth goes around the sun, different parts get more sunlight. That makes the seasons.

Astronomers also now know that our solar system is almost 5 billion years old. It is located at the edge of the Milky Way galaxy. It is one of at least 100 billion galaxies in the universe. Each one of them has billions of stars! Our galaxy alone has about 200 billion stars.

The Milky Way is a spiral galaxy. It can be seen on a very dark night as a bright band in the sky. It is so wide that light would take 100,000 years to travel across it.

Comprehension Question

People used to think that Earth was the center of everything. How were they wrong?

Our Place in Space

Have you ever looked up into the night sky and wondered what was out there? Throughout time, astronomers have gazed to the heavens, hoping to find clues about our place in the universe.

Long ago, people assumed that Earth was the center of the universe. They believed that the sun, moon, stars, and other planets orbited around it. Astronomers finally learned that Earth is one of at least eight other planets that travel around the sun.

the Sun

Our Sun

The sun is a star at the center of the solar system. It isn't the biggest or brightest star in our galaxy, but it is the star closest to Earth. It is the largest body in our solar system: in fact, it contains 99.8 percent of the solar system's mass.

Like other stars, the sun is made up of hot hydrogen and helium gases. The temperature in its center is about 15 million degrees Celsius (27 million degrees Fahrenheit). The sun's mass gives it a powerful gravitational pull. That gravity keeps Earth and other planets of the solar system in place and orbiting around the sun.

The sun, through its light and heat, is the primary source of energy on Earth. It's responsible for the growth of plants, winds, ocean currents, and the water cycle. Without the sun's energy, Earth would be very cold. It would be so cold that no living thing could survive.

The Four Seasons

Earth is tilted to one side as it orbits around the sun, and it is that tilt that causes the seasons. For part of the year, the northern hemisphere of Earth leans toward the sun and gets direct sunlight. So, it's summer there. At the same time, the southern hemisphere leans away from the sun and gets less sunlight. So, it's winter there.

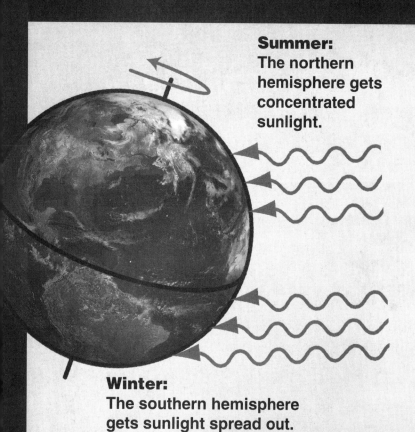

Summer: The northern hemisphere gets concentrated sunlight.

Winter: The southern hemisphere gets sunlight spread out.

The tilt doesn't change. It is the position of the planet around the sun that changes. As Earth goes around the sun, different parts get more sunlight. So at one part of the year, it is winter in the northern hemisphere of the planet. At the opposite time of the year, the Earth is tilted differently and it is summer in the northern hemisphere.

Astronomers also now know that our solar system is almost 5 billion years old. It is located at the edge of the Milky Way galaxy. It is one of at least 100 billion galaxies in the universe. Each one of them has billions of stars! Our galaxy alone has about 200 billion stars.

The Milky Way is a spiral galaxy with many arms. On very dark nights, you can see the neighboring arm as a bright band in the sky. The galaxy is so wide that light would take 100,000 years to travel across it.

Comprehension Question

Describe Earth's position relative to the rest of the universe.

Other Citizens of the Solar System

Astronomers study the stars in the sky. They try to find new things. In 1913, Percival Lowell found a new dot of light. It was very far away. It was tiny. It was hard to find. It was hard to track. The light came from two small balls of rock and ice. He named them Pluto and Charon. They were like planets. They were very small. Astronomers could not decide what to do with Pluto and Charon.

The Little Ones

The solar system was formed 4.6 billion years ago. Many things orbit the sun. They go around the sun. Some are planets. Some are smaller. They are the left-over parts. They come in lots of shapes. They come in lots of sizes. Sometimes their paths cross the orbits of planets.

Asteroids are chunks of rock. They float through space. They can be the size of whole islands. Or they can be just a few meters wide. Most are found in a space shaped like a belt. It lies between Mars and Jupiter.

Most meteoroids are very small rocks. They float in space. Some fall to Earth. They burn up while speeding through the air. When they do this they are known as meteors. They are also called shooting stars. Some large meteoroids have hit Earth's surface. They have made big craters. They are called meteorites when they hit Earth.

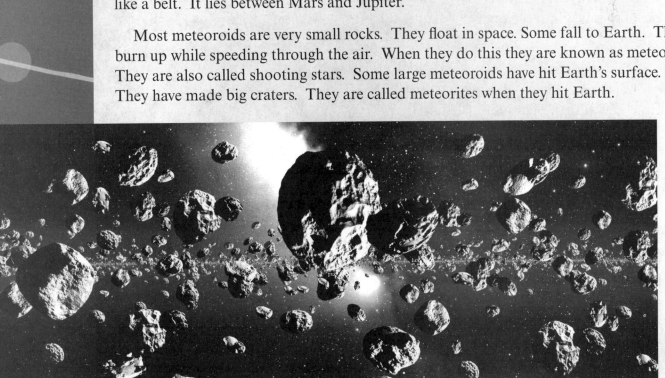

Comets are balls of dust, rock, gas, and ice. Some call them "dirty snowballs." The snowball is the comet's head. When a comet gets close to the sun, the ice melts. Then we can see a bright line of dust and gas. It can be millions of kilometers long. This is called the comet's tail. The dust and gas shine because of light from the sun.

The Big Ones

Lowell found Pluto and Charon. Others found more specks all over the sky. These specks are like planets, but smaller. They have been named things like Sedna and Quaoar. Some have codes like 2003 UB313 and 2004 DW.

Most of these specks have been found in the Kuiper Belt. That's a cloud of icy chunks shaped like a disk. It is beyond Neptune. Ceres used to count as a planet. In 1802, it was decided that Ceres was a big, round asteroid.

Scientists had to decide what counted as a planet. They had to decide what did not. A planet had to orbit the sun. It had to be round. It had to be the only thing in its orbit. Pluto's orbit crosses Neptune's path. It goes through the Kuiper Belt. There are all sorts of things in the way. Pluto can't be a planet.

Don't feel sorry for Pluto, though. It is joined by Ceres and Eris (or UB313). They are in a brand-new category called dwarf planets.

Comprehension Question

List four things in the solar system besides planets.

Other Citizens of the Solar System

Astronomers are always studying the heavens. They try to find new things. In 1913, Percival Lowell found a new speck of light. It was far off at the edge of the solar system. It was tiny. It was hard to find and track. The light came from two small balls of rock and ice. Scientists named them Pluto and Charon. They were like planets, but they were very small. Astronomers have argued for almost a hundred years about what to do with Pluto and Charon.

The Little Ones

The solar system was formed 4.6 billion years ago. There are many things that orbit the sun. Some are planets. Some are smaller. They are the left-over pieces. They come in all shapes and sizes. Sometimes their orbits cross the orbits of planets.

Asteroids are chunks of rock. They can be hundreds of kilometers across. Or they can be just a few meters wide. Most are found in a belt that lies between Mars and Jupiter.

Most meteoroids are very small rocks. They float in space. Some fall to Earth. They burn up while speeding through the air. When they do this they are known as meteors. They are also called shooting stars. Some large meteoroids have hit Earth's surface.

They have made big craters. They are called meteorites when they hit the Earth.

Comets are balls of dust, rock, gas, and ice. They are sometimes referred to as "dirty snowballs." The snowball is the comet's head. When a comet gets close to the sun, its ice melts. Then we can see a bright streak of dust and gas millions of kilometers long. This is called the comet's tail. The dust and gas shine because of sunlight.

The Big Ones

Lowell found Pluto and Charon. Others found more specks all over the sky. These objects are like planets, but smaller. They have been given names like Sedna and Quaoar. Some have been given codes like 2003 UB313 and 2004 DW.

Most of these specks have been found in the Kuiper Belt. That's a disk-shaped cloud of icy chunks. It is beyond Neptune. There's also Ceres. Astronomers used to count it as a planet. In 1802, they decided it was a big, round asteroid.

Scientists had to decide what counted as a planet and what didn't. They decided that a planet had to orbit the sun. It had to be round. It had to be the only thing in its orbit. Pluto's orbit crosses Neptune's orbit. It goes through the Kuiper Belt. There are all sorts of things in the way. Pluto can't be a planet.

Don't feel sorry for Pluto, though. It is joined by Ceres and Eris (or UB313). They are in a brand-new category called dwarf planets.

Comprehension Question

List four things that orbit the sun that are not planets.

Other Citizens of the Solar System

Astronomers are always studying the heavens, trying to discover new things. In 1913, the astronomer Percival Lowell discovered a new speck of light far off at the edge of the solar system. It was tiny, and incredibly hard to find and track. The light came from two small balls of rock and ice that scientists named Pluto and Charon. They were like planets, but they were incredibly small. Astronomers argued for decades about what to do with Pluto and Charon.

The Little Ones

There are many objects of all shapes and sizes in orbit around the sun. They were left over when the solar system was formed 4.6 billion years ago. Sometimes their orbits cross the orbits of planets.

Asteroids are chunks of rock. They can range in size from hundreds of kilometers across to just a few meters. Most are found in an asteroid belt between Mars and Jupiter.

Most meteoroids are very small rocks in space. They usually burn up while speeding through the atmosphere. Meteoroids that do this are known as meteors. They are also called shooting stars. Some large meteoroids have hit Earth's surface and made big craters. They are called meteorites.

Comets are balls of dust, rock, gas, and ice. They are sometimes referred to as "dirty snowballs." The snowball is the comet's head. When a comet gets close to the sun, its ice melts. Then we can see a bright streak of dust and gas millions of kilometers long. This is called the comet's tail. The dust and gas shine because of sunlight.

The Big Ones

Once scientists found Pluto and Charon, they started finding more specks in the sky everywhere. Planet-like objects orbiting the sun have been given names like Sedna and Quaoar. Some have been given codes like 2003 UB313 and 2004 DW.

Most of these objects have been found in the Kuiper Belt. That's a disk-shaped cloud of icy chunks beyond Neptune. There's also Ceres, which astronomers used to count as a planet until 1802. Then they decided it was just a big, round asteroid.

Scientists had to decide what counted as a planet and what didn't. They decided that a planet had to orbit the sun. It had to be round. It had to be the only thing in its orbit. Pluto's orbit crosses Neptune's orbit. It goes through the Kuiper Belt. There are all sorts of things in the way. Pluto can't be a planet.

Don't feel sorry for Pluto, though. It is joined by Ceres and Eris (or UB313). They are in a brand-new category called dwarf planets.

Comprehension Question

How are asteroids, meteoroids, comets, and dwarf planets unlike Earth?

Other Citizens of the Solar System

Astronomers are always studying the heavens, trying to discover new things. In 1913, the astronomer Percival Lowell discovered a new speck of light far off at the edge of the solar system. It was incredibly hard to find and track. The light came from two small balls of rock and ice that scientists named Pluto and Charon. They appeared to be like the inner planets, but they were incredibly small. Astronomers argued for decades about what to do with Pluto and Charon.

The Little Ones

There are many objects of all shapes and sizes in orbit around the sun. They were left over when the solar system was formed 4.6 billion years ago. Sometimes their orbits cross the orbits of planets.

Asteroids are chunks of rock, and they can range in size from hundreds of kilometers across to just a few meters. Most are found in an asteroid belt between Mars and Jupiter.

Most meteoroids are also chunks of rock, smaller than asteroids. When they fall into Earth's gravity, they burn up speeding through the atmosphere. Scientists call them meteors, but you may be familiar with them as shooting stars. Large meteoroids impact Earth's surface and make big craters; their remains are called meteorites.

Comets are balls of dust, rock, and ice, which is why they are sometimes referred to as "dirty snowballs." The snowball is the comet's head, and when a comet gets close to the sun, its ice melts. The melted ice streaming off the head creates a bright streak of dust and gas millions of kilometers long called the comet's tail. The dust and gas shine because they reflect sunlight.

The Big Ones

Once scientists found Pluto and Charon, they started finding more specks in the sky everywhere. Planet-like objects orbiting the sun have been given names like Sedna and Quaoar. Some have been given codes like 2003 UB313 and 2004 DW.

Most of these objects have been found in the Kuiper Belt, a disk-shaped cloud of icy chunks beyond Neptune. There's also Ceres, which astronomers used to count as a planet. In 1802, they decided it was just a big, round asteroid.

Scientists had to decide what counted as a planet and what didn't. They decided that a planet had to orbit the sun, had to be round, and had to be the only astronomical body in its orbit. Pluto's orbit crosses Neptune's orbit and careens through the Kuiper Belt, so there are all sorts of things in its orbit. Pluto can't be a planet.

Don't feel sorry for Pluto, though. Together with Ceres and Eris (or UB313), it goes into a brand-new category: dwarf planets.

Comprehension Question

Contrast the orbiting bodies described in this passage with planets such as Earth.

The Astronomer's Toolbox

Have you ever looked up at the stars? Have you thought about what is up there? Is there life? How did it all begin? If you asked these things, you are not alone! Astronomers are scientists who study the stars. They try to figure out how the universe works.

Long ago astronomers did not have many tools. They used their own eyes. They built observatories on the ground. They looked up into the sky. Today, we have tools in space. They orbit Earth. With these new tools, we can learn a great deal.

The Great Observatories

Today we have telescopes. They give us a good view of the stars. Even telescopes have limits, though. Earth's air is made of gas and dust. It blocks our view. So NASA launched four telescopes into space. They are called NASA's Great Observatories.

They give us a much better view. They take pictures of planets and stars. Some pictures are even of Earth! They send them back to us. Now we can study space from home!

The Hubble Space Telescope was the first of its kind. It is also the best known. It takes pictures of the light we see. It sees what we would see if we were up there. It was launched in 1990. It flies more than 600 kilometers (nearly 400 miles) above Earth. It is run from the ground. The pictures it has sent to us are pretty. They help us learn about many things. Scientists learn about stars' births and deaths. They even learn how the galaxy was formed.

The Compton Gamma Ray Observatory was next. It was launched in 1991. It went up in a space shuttle. It weighed three thousand pounds! Compton studied things called gamma rays. Gamma rays are a kind of

light. Our eyes alone cannot see them. It made maps of the sky. It was up there for nine years. Then a part broke. NASA did not want it to fall. They brought it down. It came down in the Pacific Ocean in 2000.

The third satellite is the Chandra X-Ray Observatory. Chandra's mirrors are great. They are the largest and smoothest ever built. It sees through clouds of gas. They are called nebulae. This telescope picks up X-rays. They come from black holes. Three students used data from Chandra. They found a neutron star!

NASA's Spitzer Space Telescope is the last one. It was launched in 2003. It sees infrared light. That way it sees through clouds of dust in space. This helps us find young stars. We can find far off planets. Spitzer is named for an astronomer. It was his idea to put telescopes in space.

Comprehension Question

Where are the Great Observatories?

The Astronomer's Toolbox

Have you ever gazed up at the stars? Have you ever thought there might be life there? Have you ever thought about how it all began? If you answered yes to any of these questions, you are not alone! Astronomers are scientists. They study the heavens. They try to figure out how the universe works.

Early astronomers did not have many tools. They used their own eyes. They built observatories on the ground. They looked up into the heavens. Today, we have tools in orbit around Earth. With these tools, we can learn a great deal.

The Great Observatories

Modern observatories use telescopes. They give us a good view of the stars. Even today's telescopes have limits, though. Earth's air is made of gas and dust. It blocks our view. So NASA launched four

telescopes into space. They are called NASA's Great Observatories. They give us a much better view. They take pictures and send them back to us. The pictures are of planets, stars, and even Earth. Scientists can study outer space without leaving home!

The Hubble Space Telescope was the first. It is also the best known. It takes pictures of visible light. It sees what we would see if we were in orbit. It was launched in 1990. It orbits more than 600 kilometers (nearly 400 miles) above Earth's surface. Scientists on the ground control it. The pictures it has sent help them learn about many things. They learn about stars' births and deaths. They even learn how the galaxy was formed.

The Compton Gamma Ray Observatory was next. It was launched in 1991. It went up in the space shuttle *Atlantis*. It weighed 17 tons! Compton studied gamma rays. Gamma rays are a kind of light that our eyes

alone cannot see. It made maps of the sky. It was in orbit for nine years. Then a part broke. NASA did not want it to fall. They decided to bring it out of orbit. It came down in the Pacific Ocean in 2000.

The third satellite is the Chandra X-Ray Observatory. Chandra's mirrors are the largest and smoothest ever built. It sees through clouds of gas. They are called nebulae. This telescope picks up X-rays. They come from black holes. Three students used images from Chandra. They found a neutron star!

NASA's Spitzer Space Telescope is the last piece of the program. It was launched in 2003. It senses infrared light. That way it sees through clouds of dust in space. This helps scientists find young stars. They can find new solar systems. Spitzer is named for an astronomer. He suggested placing telescopes in space.

Comprehension Question

Why are the Great Observatories in orbit?

The Astronomer's Toolbox

Have you ever gazed up at the stars? Have you ever wondered if there might be life on other planets? Have you ever wondered how the universe began? If you answered yes to any of these questions, you are not alone! Astronomers are scientists who study the heavens. They try to discover the mysteries of the universe.

Ancient astronomers did not have many tools. They used their own eyes. They built observatories on the ground to look up into the heavens. Today, astronomers have tools in orbit around Earth. With these tools, we can learn a great deal.

The Great Observatories

Modern observatories use telescopes. They give us an amazing view of the heavens. Even today's observatories have limits, though. Earth's atmosphere is made of gas and dust. It blocks our view.

So NASA launched four telescopes into space. They are called NASA's Great Observatories. They give us a much better look. They take and send back to us pictures of planets, stars, and even Earth. Scientists can study outer space without leaving home!

The Hubble Space Telescope was the first. It is also probably the best known. It takes pictures of visible light. It sees what we would see if we were in orbit. It was launched in 1990. It orbits more than 600 kilometers (nearly 400 miles) above Earth's surface. Scientists on the ground control it. The pictures it has sent help them understand many things. They learn about star birth, star death, and even how the galaxy was formed.

The Compton Gamma Ray Observatory was next. It was launched from the space shuttle *Atlantis* in 1991. It weighed 17 tons! The satellite studied gamma rays.

Gamma rays are a kind of light invisible to the naked eye. It made maps of the sky. It was in orbit for nine years. Then a gyroscope broke. NASA decided to bring it out of orbit instead of letting it fall. It came down in the Pacific Ocean in 2000.

The third satellite is the Chandra X-Ray Observatory. Chandra's mirrors are the largest and smoothest ever built. This telescope picks up the energy that comes from black holes. It sees through clouds of gas called nebulae. Three students used images from Chandra to find a neutron star!

NASA's Spitzer Space Telescope is the last piece of the program. It was launched in 2003. It senses infrared light. This helps astronomers see through clouds of dust in space. This helps scientists find young stars and new solar systems. Spitzer is named for the first scientist to suggest placing telescopes in space.

Comprehension Question

What are the benefits of observatories in space?

The Astronomer's Toolbox

Have you ever gazed up at the heavens and wondered if there might be life on other planets, or how the universe began? If you answered yes to any of these questions, you are not alone! Astronomers are scientists who study the heavens, trying to discover the mysteries of the universe.

Ancient astronomers did not have many tools. They used their own eyes. They built observatories on the ground to look up into the heavens. Today, astronomers have tools in orbit around Earth. With these tools, we can learn a great deal.

The Great Observatories

Modern observatories use telescopes to give us an amazing view of the heavens. Even today's observatories have limits, though. Earth's atmosphere of gas and dust blocks our view. So NASA launched four telescopes into space in the Great Observatories Program. They take and send back to us pictures of planets, stars, and even Earth without the interference of our atmosphere. Scientists can study outer space without leaving home!

The Hubble Space Telescope was the first, and is probably the best known. It takes pictures of visible light: what we would see if we were in orbit. It was launched in 1990. It orbits more than 600 kilometers (nearly 400 miles) above Earth's surface, but scientists on the ground control it. The pictures it has sent help them understand star birth, star death, and even how the galaxy was formed.

The Compton Gamma Ray Observatory was launched from the space shuttle Atlantis in 1991. It weighed 17 tons! The satellite studied gamma rays, a kind of light invisible to the naked eye. For nine years, it made maps of the sky from orbit. Then a

gyroscope broke, and NASA decided to bring it out of orbit instead of letting it fall. It came down in the Pacific Ocean in 2000.

Chandra X-Ray Observatory, the third satellite in the program, possesses the largest and smoothest mirrors ever built. This telescope picks up the energy that comes from black holes. It can see through clouds of gas called nebulae. Three students used images from Chandra to find a neutron star!

NASA's Spitzer Space Telescope, launched in 2003, is the last piece of the program. By sensing infrared light, Spitzer allows astronomers to see through clouds of dust in space. This helps scientists find young stars and new solar systems. Spitzer is named for the first scientist to suggest placing telescopes in space.

Comprehension Question

What do you think scientists hoped to accomplish by launching observatories into space?

The Journey to Space

The space age got its start in 1957. The Soviet Union put *Sputnik 1* into space. It was the first man-made thing in orbit. Four years later there was a man named Yuri Gagarin. He did one more "first." He was the first man to fly in space.

Today, the largest space group is in the U.S. It is called NASA. It launched the *Apollo 11* space capsule. It was in 1969. It gave the United States its own "first." Neil Armstrong was on *Apollo 11*. He was the first person to walk on the moon. His first words there are well known. He said, "One small step for man, one giant leap for mankind."

NASA then built the space shuttle. That is a kind of spacecraft. It can be used over and over. Shuttles have flown since 1981. They have had more than one hundred flights. Sadly, there have been accidents. Two shuttle crews have been lost.

In 1983, there was a space probe called *Pioneer 10*. It had its own "first." It was the first man-made thing to leave the solar system. It took 11 years to get there!

Our Base in Space

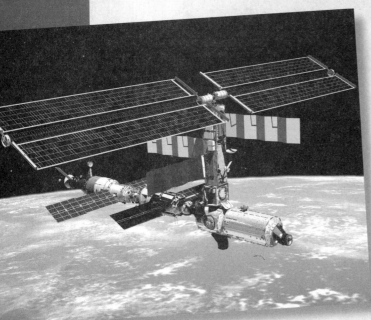

Did you know there is a lab up in space? It is called the ISS. It is a bright tiny spot in the night sky. It is like a home in space. Astronauts live there. They do experiments. People from sixteen nations have worked on it. There are at least two people on board at all times. The first crew got there in 2000. Most crew members stay about six months.

The crew does special experiments. They can only be done in space. They also can see stars from above Earth's air. Scientists on land support the ISS crew. They watch their health. They help with experiments. One day, the ISS may serve as a launch pad. A trip to Mars can start there.

The Future of Space

What does the future of space hold for us? The next trip to the moon is planned for 2018. How old will you be in 2018? The trip will last a week. The astronauts will make their own water. They will make their own fuel. They will learn how to live on the moon. There are also plans to visit Mars. This is planned for 2028. This trip would last much longer. Astronauts could be there for 500 days!

Pluto should get a visit from Earth, too. Pluto is a dwarf planet. It is very far away. In January 2006, NASA launched New Horizons. It is a probe. It went on a long trip. It will reach Pluto in 2015. The probe will fly by Pluto. It will take pictures. It will send data to Earth. The probe is sure to find new things. They will help us learn.

Comprehension Question

List three ways people have visited space.

The Journey to Space

The space age got its start in 1957. The Soviet Union launched *Sputnik 1*. It was the first man-made satellite. Four years later, a man named Yuri Gagarin did one more "first." He was the first person to fly into space.

Today, the largest space research group is in the U.S. It is the National Aeronautics and Space Administration (NASA). Its Apollo 11 mission gave the United States its own "first." It was in 1969. Neil Armstrong was the first person to walk on the moon. His first words there are famous. He said, "One small step for man, one giant leap for mankind."

NASA then built the space shuttle. That is a kind of spacecraft. It can be used over and over. The shuttles have flown since 1981. They have had more than 100 flights. Sadly, two shuttle crews have been lost in accidents.

In 1983, the space probe *Pioneer 10* had its own "first." It was the first man-made object to leave the solar system. It took 11 years to get there!

Our Base in Space

Did you know there is a lab up in space? It is called the International Space Station, or ISS. It is a bright speck in the night sky. It is like a home in space. Astronauts live there and do experiments. People from sixteen nations have worked on it. There are always at least two people on board. The first crew got there in 2000. Most crew members stay about six months.

The crew does special experiments. They can only be done in space. They also can see stars from above the Earth's air. Scientists on land support the ISS crew. They watch their health. They help with experiments. One day, the ISS may serve as a launch pad. A trip to Mars can start there.

The Future of Space

What does the future hold for exploring space? The next trip to the moon is planned for 2018. How old will you be in 2018? The trip will last seven days. The astronauts will make their own water and fuel. There are also plans to visit Mars by 2028. This would last much longer. Astronauts could be there for 500 days!

Pluto should get a visit from Earth, too. Pluto is a dwarf planet. It is at the edge of the solar system. In January 2006, NASA launched New Horizons. It is a probe. It went on a long trip. It will reach Pluto in 2015. The probe will fly by Pluto. It will send pictures and data to Earth. The probe is sure to find surprises. They will help us learn more about our solar system and our universe.

Comprehension Question

Describe the goals of three different space missions from the passage.

The Journey to Space

The space age began in 1957. The Soviet Union launched *Sputnik 1*. It was the first man-made satellite. Four years later, a man named Yuri Gagarin did one more "first." He became the first person to pilot a spacecraft.

Today, the largest space research group is in the U.S. It is the National Aeronautics and Space Administration (NASA). Its Apollo 11 mission gave the United States its own "first." It was in 1969 with astronaut Neil Armstrong. He became the first person to walk on the moon. His first words there are famous. He said, "One small step for man, one giant leap for mankind."

NASA later developed the space shuttle. That is a spacecraft that can be used over and over. The shuttles have flown since 1981. They have had more than 100 flights. Sadly, they have lost two crews in tragic accidents.

In 1983, the space probe *Pioneer 10* had its own "first." It became the first man-made object to leave the solar system. It was launched 11 years earlier!

Our Base in Space

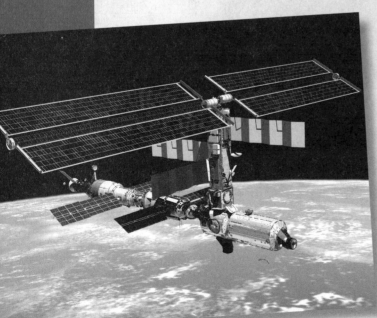

Did you know there is a laboratory up in outer space? It is called the International Space Station, or ISS. It is one of the brightest objects in the night sky. It is like a home in space. Astronauts live there and carry out experiments. Sixteen nations have worked together on it. There are always at least two people on board. The first crew arrived in 2000. Most crew members stay about six months.

The crew does special experiments. They can only be done in space. They also observe the universe from outside the atmosphere.

Scientists on land are always in support of the ISS crew. They monitor their health and help with experiments. One day, the ISS may serve as a launch pad for missions to other planets such as Mars.

The Future of Space

What does the future hold for space exploration? NASA's plans for the future will take us to new heights! The next manned trip to the moon is planned for 2018. This mission will last about seven days. Scientists want astronauts to be able to produce water, fuel, and other necessities for life. Can homes on the moon be far behind? There are also plans for astronauts to visit Mars by 2028. This would be a much longer mission. Astronauts could be on the planet's surface for 500 days!

Pluto should be getting a visit from Earth, too. Pluto is a dwarf planet. It is at the edge of the solar system. In January 2006, NASA launched the *New Horizons* spacecraft. It began a very long trip. It will reach Pluto in 2015. The unmanned spacecraft will fly by Pluto. It will send images and data to Earth. New Horizons is sure to find surprises. They will help us learn more about our solar system and our universe.

Comprehension Question

List three different reasons people have sent missions into space.

The Journey to Space

The space age began in 1957 when the Soviet Union launched *Sputnik 1*. This was the world's first man-made satellite. Four years later, Soviet cosmonaut Yuri Gagarin became the first person to pilot a spacecraft.

Today, the largest space research group is the U.S. National Aeronautics and Space Administration (NASA). Its Apollo 11 mission made the United States the first country to put a person on the moon. In 1969, astronaut Neil Armstrong became the first person to walk on the moon. His first words there are famous. He said, "One small step for man, one giant leap for mankind."

NASA later developed the space shuttle: a spacecraft that can be used over and over. Since 1981, the space shuttle fleet has had more than 100 successful flights. Sadly, two crews were lost in tragic accidents.

In 1983, the space probe *Pioneer 10* became the first man-made object to leave the solar system. It was launched 11 years earlier!

Our Base in Space

Did you know there is a laboratory floating in outer space? In fact, the International Space Station, or ISS, is one of the brightest objects in the night sky. It is like a home in space: astronauts live there and carry out experiments. Sixteen nations have worked together on this project. There have been at least two people on board the ISS since the first crew arrived in 2000. Most crew members stay about six months.

The astronauts perform experiments that can only be done in the freefall environment of space. They also observe the universe from outside the atmosphere. Of course, land-based scientists are always in support of the ISS crew. They monitor their health and help with experiments. One day, the ISS may serve as a launch pad for missions to other planets such as Mars.

The Future of Space

What does the future hold for space exploration? NASA's plans for the future will take us to new heights! The next manned trip to the moon is planned for 2018. This mission will last about seven days. Scientists want astronauts to be able to produce water, fuel, and other necessities for life. Can homes on the moon be far behind? There are also plans for astronauts to visit Mars by 2028. This would be a much longer mission. Astronauts could be on the planet's surface for 500 days!

Pluto should be getting a visit from Earth, too. Pluto is a dwarf planet at the edge of the solar system. In January 2006, NASA's *New Horizons* spacecraft began the very long journey to the dwarf planet. It will reach Pluto in 2015. The unmanned spacecraft will fly by Pluto and send images and data to Earth. *New Horizons* may uncover surprises that will help us learn more about our solar system and our universe.

Comprehension Question

Why do we explore space?

Resources

Works Cited

August, Diane and Timothy Shanahan (Eds). (2006). *Developing literacy in second-language learners: Report of the National Literacy Panel on language-minority children and youth.* Mahwah, NJ: Lawrence Erlbaum Associates, Inc.

Marzano, Robert, Debra Pickering, and Jane Pollock. (2001). *Classroom instruction that works.* Alexandria, VA: Association for Supervision and Curriculum Development.

Tomlinson, Carol Ann. (2000). Leadership for Differentiating Schools and Classrooms, Alexandria, VA: Association for Supervision and Curriculum Development.

Image Sources

Passage	Description	Source	Filename
Trade Winds and Jet Streams	Trade Winds	Teacher Created Materials	tradewinds.jpg
Trade Winds and Jet Streams	World from Space	Photo Researchers (SC4505)	worldfromspace.jpg
Trade Winds and Jet Streams	jet streams	Teacher Created Materials	jetstreams.jpg
Water Cycle	Water Cycle Diagram	Teacher Created Materials	watercycle.jpg
Water Cycle	clouds	Alamy (AA3KMA)	clouds.jpg
Water Cycle	waves	Shutterstock (19040601)	waves.jpg
Tornadoes and Hurricanes	Tornado	Alamy (A64203)	tornado.jpg
Tornadoes and Hurricanes	Tornado Genesis Diagram	Teacher Created Materials	tornadodiagram.jpg
Tornadoes and Hurricanes	hurricane	NASA	hurricane.jpg
Structure of the Earth	Earth structure diagram	Photo Researchers (SB5877)	earthstructure.jpg
Structure of the Earth	geologists	Jason Bodine	geologists.jpg
Structure of the Earth	lava	Alamy (AF7RF7)	lava.jpg
Earthquakes and Volcanoes	Pinnacle National Monument	Shutterstock (467359)	pinnacle.jpg
Earthquakes and Volcanoes	Fault line diagrams	Teacher Created Materials	faultlines.jpg
Earthquakes and Volcanoes	Volcano diagram	Photo Researchers (SB5837)	volcano.jpg
Plate Tectonics	the tectonic plate of Earth	USGS	tectonicplates.jpg

Resources (cont.)

Image Sources (cont.)

Passage	Description	Source	Filename
Plate Tectonics	plate boundary diagram	Photo Researchers (SES853)	plateboundaries.jpg
Wegener Solves a Puzzle	Alfred Wegener	Getty Images (56461968)	wegener.jpg
Wegener Solves a Puzzle	Pangea	Teacher Created Materials	pangea.jpg
Wegener Solves a Puzzle	Present Day Continents	Teacher Created Materials	presentday.jpg
The Rock Cycle	magma diagram	Teacher Created Materials	magma.jpg
The Rock Cycle	sediment diagram	Teacher Created Materials	sediment.jpg
The Rock Cycle	banded gneiss	USGS	gneiss.jpg
Fun with Fossils	leaf fossil	Shutterstock (1057663)	leaffossil.jpg
Fun with Fossils	fish fossil	Shutterstock (1832306)	fishfossil.jpg
Fun with Fossils	dinosaur fossil	Shutterstock (1008707)	dinosaurfossil.jpg
The Inner Planets	solar system diagram	Teacher Created Materials	solarsystem.jpg
The Inner Planets	Venus	Alamy (A17J6C)	venus.jpg
The Inner Planets	Mars rover	iStockphoto.com (164525)	marsrover.jpg
The Inner Planets	Mars	Getty Images (57015794)	mars.jpg
The Outer Planets	Saturn	Photos.com (16472084)	saturn.jpg
The Outer Planets	Uranus	Photos.com (26811793)	uranus.jpg
The Outer Planets	Neptune	NASA	neptune.jpg
The Outer Planets	Lowell Observatory	Shutterstock (67903)	lowellobservatory.jpg
Our Place in Space	the sun	Shutterstock (3500010)	sun.jpg
Our Place in Space	seasonal diagram	Teacher Created Materials	seasons.jpg
Our Place in Space	Milky Way	NASA	milkyway.jpg
Other Citizens of the Solar System	Quaoar	Photo Researchers (SF4212)	quaoar.jpg
Other Citizens of the Solar System	Asteroid Belt	Photo Researchers (SE5403)	asteroids.jpg

Resources (cont.)

Image Sources (cont.)

Passage	Description	Source	Filename
Other Citizens of the Solar System	comet	Public Domain	comet.jpg
Other Citizens of the Solar System	Pluto	Photo Researchers (SF9688)	pluto.jpg
The Astronomer's Toolbox	Ptomely	Alamy (A2FJS4)	ptomely.jpg
The Astronomer's Toolbox	Hubble Space Telescope	NASA	hubble.jpg
The Astronomer's Toolbox	Spitzer Space Telescope	NASA	spitzer.jpg
The Journey to Space	moon landing	NASA	moonlanding.jpg
The Journey to Space	International Space Station	NASA	iss.jpg
The Journey to Space	Mars mission	NASA	marsmission.jpg

Resources (cont.)

Contents of Teacher Resource CD

PDF Files

The full-color pdfs provided are each eight pages long and contain all four levels of a reading passage. For example, the Jet Streams and Trade Winds PDF (pages 21–28) is the *jetstreams.pdf* file.

Text Files

The text files include the text for all four levels of each reading passage. For example, the Jet Streams and Trade Winds text (pages 21–28) is the *jetstreams.doc* file.

Text Title	Text File	PDF
Jet Streams and Trade Winds	jetstreams.doc	jetstreams.pdf
The Water Cycle	watercycle.doc	watercycle.pdf
Tornadoes and Hurricanes	tornadoes.doc	tornadoes.pdf
Structure of the Earth	structureofearth.doc	structureofearth.pdf
Earthquakes and Volcanoes	earthquakes.doc	earthquakes.pdf
Plate Tectonics	platetectonics.doc	platetectonics.pdf
Wegener Solves a Puzzle	wegener.doc	wegener.pdf
The Rock Cycle	rockcycle.doc	rockcycle.pdf
Fun with Fossils	fossils.doc	fossils.pdf
The Inner Planets	innerplanets.doc	innerplanets.pdf
The Outer Planets	outerplanets.doc	outerplanets.pdf
Our Place in Space	ourplace.doc	ourplace.pdf
Other Citizens of the Solar System	othercitizens.doc	othercitizens.pdf
The Astronomer's Toolbox	astronomerstoolbox.doc	astronomerstoolbox.pdf
The Journey to Space	journeytospace.doc	journeytospace.pdf

JPEG Files

The images found throughout the book are also provided on the Teacher Resource CD. See pages 141–143 for image descriptions, credits, and filenames.

Teacher Resource CD

Word Documents of Texts
- Change leveling further for individual students.
- Separate text and images for students who need additional help decoding the text.
- Resize the text for visually impaired students.

Full-Color PDFs of Texts
- Create overheads.
- Project texts for whole-class review.
- Read texts online.
- Email texts to parents or students at home.

JPEGs of Primary Sources
- Recreate cards at more levels for individual students.
- Use primary sources to spark interest or assess comprehension.